Debates in Science Education

Debates in Science Education explores the major issues all science teachers encounter in their daily professional lives. It encourages critical reflection and aims to stimulate both novice and experienced teachers to think more deeply about their practice, and link research and evidence to what they have observed in schools.

Written by expert science educators, chapters tackle established and contemporary issues enabling you to reach informed judgements and argue your point of view with deeper theoretical knowledge and understanding. Each chapter is supported and extended by carefully selected further reading and reflective questions.

Key debates include:

- the impact of policy on science education;
- transition from primary to secondary school;
- getting right the secondary science curriculum;
- girls in science;
- sex education and science;
- school science and technology;
- language and communication in the classroom;
- world science, local science.

With its combination of expert opinion and fresh insight, *Debates in Science Education* is the ideal companion for any student or practising teacher engaged in initial training, continuing professional development and Masters level study.

Mike Watts is Professor of Education in the School of Sport and Education, Brunel University, UK.

Debates in Subject Teaching Series

Edited by Susan Capel, Jon Davison, James Arthur and John Moss

The **Debates in Subject Teaching Series** is a sequel to the popular **Issues in Subject Teaching Series**, originally published by Routledge between 1999 and 2003. Each title presents high-quality material, specially commissioned to stimulate teachers engaged in initial training, continuing professional development and Masters level study to think more deeply about their practice, and link research and evidence to what they have observed in schools. By providing up-to-date, comprehensive coverage, the titles in the **Debates in Subject Teaching Series** support teachers in reaching their own informed judgements, enabling them to discuss and argue their point of view with deeper theoretical knowledge and understanding.

Debates in History Teaching
Edited by Ian Davies

Debates in English Teaching
Edited by Jon Davison, Caroline Daly and John Moss

Debates in Religious Education
Edited by Philip Barnes

Debates in Citizenship Education
Edited by James Arthur and Hilary Cremin

Debates in Art and Design Education
Edited by Lesley Burgess and Nicholas Addison

Debates in Music Teaching
Edited by Chris Philpott and Gary Spruce

Debates in Physical Education
Edited by Susan Capel and Margaret Whitehead

Debates in Geography Education
Edited by David Lambert and Mark Jones

Debates in Mathematics Education
Edited by Dawn Leslie and Heather Mendick

Debates in Science Education

Edited by
Mike Watts

Routledge
Taylor & Francis Group

LONDON AND NEW YORK

First published 2014
by Routledge
2 Park Square, Milton Park, Abingdon, Oxon OX14 4RN

and by Routledge
711 Third Avenue, New York, NY 10017

Routledge is an imprint of the Taylor & Francis Group, an informa business

© 2014 Mike Watts

The right of the editor to be identified as the author of the editorial material, and of the authors for their individual chapters, has been asserted in accordance with sections 77 and 78 of the Copyright, Designs and Patents Act 1988.

British Library Cataloguing in Publication Data
A catalogue record for this book is available from the British Library

Library of Congress Cataloging in Publication Data
A catalog record for this book has been requested

ISBN: 978-0-415-65828-7 (hbk)
ISBN: 978-0-415-65829-4 (pbk)
ISBN: 978-1-315-86756-4 (ebk)

Typeset in Galliard
by Book Now Ltd, London

Printed and bound in the United States of America by
Edwards Brothers Malloy

Contents

Contributors

Michael Allen: After a first career in the Royal Air Force, Dr Michael Allen taught science in Berkshire secondary and middle schools for twelve years. He now lectures full-time in primary science education at Brunel University, London. His research interests have included the elicitation and correction of science misconceptions by means of constructivist philosophy-driven conceptual change pedagogies, confirmation bias and its influence on science observation/inference making (in both historical and contemporary contexts), and the use of practical work to enhance achievement in school science. Lately his research has taken him into the nursery, exploring the biological ideas of pre-school children with a view to illuminating the early origins of misconceptions.

Steve Alsop is a Professor in the Faculty of Education and Department of Science and Technologies Studies at York University, Canada. Steve has held a series of academic positions including associate dean, academic director, coordinator, honorary professor, secondary and primary school teacher. He has a wide range of research interests embracing (and embodying) the social and political organisation of scientific and technological knowledge, pedagogies of science and technology, and environmental sustainability. He lives in Toronto.

David Barlex is an acknowledged leader in design and technology education, curriculum design and curriculum materials development. He taught science and technology in comprehensive schools for fifteen years before becoming a teacher educator. He directed the Nuffield Design and Technology Project, and was Educational Manager of Young Foresight, an initiative that has developed approaches to teaching and learning to enhance students' ability to respond creatively to design and technology activities. In 2002 he won the DATA Outstanding Contribution to Design and Technology Education award. David's research activity stems from his conviction that there should be a dynamic and synergic relationship between curriculum development and academic research. He currently pursues this activity through partnerships with researchers in the UK, Canada, Australia and the USA. He has presented work regularly at international conferences and published in the *Journal of Design and Technology* and the *International Journal of Technology and Design Education Research*. Most recently he has acted as the STEM (science, technology, engineering and mathematics) consultant and a curriculum adviser for the Digital Design and Technology project to the Design and Technology Association.

Joanne Cole is a lecturer in Electronic and Computer Engineering at Brunel University and currently acts as the first year coordinator for all Engineering programmes within her subject area. She is also an active experimental particle physicist, having been a member of the Compact Muon Solenoid (CMS) experiment on the Large Hadron Collider at CERN in Geneva, Switzerland for seven years. She completed her doctorate whilst at Imperial College working on the ZEUS experiment at DESY in Hamburg, Germany. Within the CMS experiment she has been involved in the final phase of construction and commissioning of the detector and has focussed on studying the heaviest of the known fundamental particles – the top quark – using data collected by CMS. In addition to her research, Joanne is an active member of the Institute of Physics, having recently reached the end of her term as Honorary Secretary of the Nuclear and Particle Physics Division. She is also heavily involved in the promotion of women in science at all ages: She is currently the principal organiser of a series of outreach events aimed at encouraging girls to study physics, and is a member of the Brunel University Athena SWAN self-assessment team.

Nic Crowe is a qualified teacher, youth and community worker, and Fellow of The Higher Education Academy. He is an experienced practitioner with a background in technology and cultural studies. His main research focuses on the fantasy cultures of youth, specifically digital technology, fantasy role play and comics, anime and manga. His research explores the links between digital play and the curriculum, in particular the ways that digital games can be used as tools for learning. Nic is currently writing about the 'dark side' of digital technology' related to 'Internet trolls', 'grief tourism' and 'Pro-Ana/Pro-Mia digital presences'. He has made a number media appearances talking about on-line gaming and its associated cultures, including a presentation of his Runescape research for Radio 4's programme, *Thinking Allowed*.

John K. Gilbert is Professor Emeritus at University of Reading, and visiting professor at King's College London. He has spent the last thirty years researching, thinking and writing about some of the key and enduring issues in science education. He was awarded a Lifetime Achievement Award for his contribution to science teaching in 2001 by the American National Association of Research into Science Teaching. He has contributed over twenty books and 400 articles to the field and is Editor-in-Chief of the *International Journal of Science Education*.

Peeranut Kanhadilok is Science Educator for the exhibition section of the National Science Museum, Bangkok, Thailand. Over a ten-year period she has been responsible for the museum's public activity programmes, exhibitions and museum research, with emphasis on developing science activities, science and technology exhibitions – particular within the traditional technology gallery. Her research explores the learning of both science and local Thai wisdom through the museum's activities and exhibitions. She has developed and researched a variety of science activities, for example scientific toys, traditional Thai toys, science cultural camps, science shows, science drama and many science events at the National Science Museum. In 2011 she was awarded a Museum scholarship, under the auspices of the Thai Government, to undertake doctoral study at Brunel University, London. She has conducted a major study of family play-learning through make-and-play activities with traditional Thai toys. Through her chapter in this book, journal articles and many conference papers, Peeranut has explored family learning in science museums, and the 'bi-gnostic learning' of western science and Thai local wisdom.

Sara Leite has a BSc in Physics (minor in Chemistry) and an MSc in Teaching of Physics and Chemistry from the University of Aveiro, Portugal. While studying for her BSc she initiated research in physics and, while a trainee teacher of physics and chemistry, she was engaged in classroom-based research focused on student-centred approaches to teaching. She teaches chemistry at a secondary school in Aveiro.

Chris Lloyd-Staples graduated in Natural Sciences at Cambridge and went straight into teaching. After twenty years as a science teacher and head of department in comprehensive schools, Chris went to work in Buckinghamshire as the Science Adviser for the county, at a time that coincided with implementation of the National Strategy. He worked with primary and secondary schools, supporting teachers as they adopted, and then largely abandoned, the National Strategy messages. This unique insight into the implementation of a major top-down initiative and its subsequent dilution formed the basis of a doctorate at Brunel University. The failure of such a flagship national undertaking has left a profound concern that well-intended and valuable changes in teaching can founder if the method of introduction is not better designed. After thirteen years working in school improvement, Chris now has responsibility for advising on curriculum within Buckinghamshire, and for the close support of a number of schools. Re-organisation at national and local levels, alongside changing policies from the DfE, is making this a very uncertain time, but Chris believes that specialist support for schools will remain vitally important in securing good outcomes for young people. Chris is married with two children, and teaching is clearly in the genes – the whole family are involved in education. For relaxation, Chris is a model-maker and writer for hobby magazines.

Richard Malthouse is a Teaching Fellow at Brunel University where he lectures in Contemporary Education. His interests lie in reflective practice and in particular the sociological aspects of Situated Reflective Practice. Richard works as a consultant for the Institute for Learning and is a Fellow of that organisation.

Judith Miller is a senior lecturer in the School of Education, University of New England, Armidale, New South Wales, Australia, with over twenty years' experience in pre-service teacher education, specialising in Health and Physical Education K-12. Her research focuses on the health of young people, particularly in rural contexts. The issue of how to prepare primary generalists to teach sex education has been a long-standing challenge to the limited pre-service teacher education space within which Judith and other academics operate in Australia. Based on international contacts and shared conversations on this issue, this challenge is not isolated to UNE or Australia. Judith has recently co-authored a book preparing future educators for the Australian Curriculum in Health and Physical Education, and has school-based teaching experience in the primary to secondary levels of schooling for health and physical education in the USA and Australia.

Helena Pedrosa-de-Jesus is an Associate Professor with Agregação, in the Department of Education, University of Aveiro, Portugal. She is an active researcher at the Research Centre 'Didactics and Technology in Education of Trainers', and her main research interests are in learners' questioning in education, and in school supervision of trainee teachers of science. Her special emphasis has been on the role of questions in reflecting on learning and teaching, towards academic development. Over the last twenty-five years, this field of work has been at the heart of her research activity, working inside classrooms in higher education, secondary and primary schools. She has published widely in the field of science education and has been very active in structuring and coordinating an international network devoted to discussing and exchanging experiences and research results on questions and questioning used in classroom teaching and learning.

Frances Quinn has worked as a scientist and science educator for close to twenty-five years, in secondary and higher education sectors. After teaching secondary school science, and biological science and ecology at the University of New England, Australia, she proceeded to a research and teaching focus in Science Education. Her research interests span students' perceptions of learning science, socio-scientific issues in science education, Education for Sustainability and teaching/learning in science.

Susan Rodrigues is currently Professor in Education at the University of Northumbria, England. Her previous academic experiences have been with the University of Dundee, Scotland; University of Stirling, Scotland; University of Durham, England; and the University of Melbourne, Australia. She conducts research into the teaching and learning of science, where her work is making a contribution both to innovative practice and to theory. Her research explores issues such as the language of science, teacher professional development, and how ICT can be used to transform teaching, learning and assessment in science. Susan is known internationally for her science education research, and has been invited to present at international events, including in Australia, Brunei, Germany, Poland and Cyprus. She was a guest editor for the *Science Education International* journal and is referee for several international science education and technology education journals. She is currently working on projects funded by Astrazeneca Science Teaching Trust and is working in partnership with the European Union F7 programme.

Jodi Roffey-Barentsen is Programme Manager of Education and Teacher Training at Farnborough College of Technology. She is responsible for the development and delivery of a suite of degree courses in education and learning support. Her research interests include reflective practice, the deployment of Teaching Assistants, and fransition from further to higher education. Jodi works as a consultant for an international exam board and is a Fellow of the Institute for Learning.

Marion Rutland is currently an Honorary Research Fellow at the University of Roehampton, London, where she is the Design and Technology Course Leader for school-based graduate teachers for the West London Programme, MA Tutor and a PhD supervisor. Other roles at the University have included Principal Lecturer and Course Leader for the PGCE Design and Technology Secondary programme and Curriculum Leader for the primary Design and Technology programmes. Prior to this she taught for twenty years in a range of secondary schools, and was an Advisory Teacher for ICT in London. Her search interests include teaching and learning in design and technology, creativity and teaching food technology in schools.

Keith S. Taber is Reader in Science Education at the University of Cambridge. He is Chair of the Science, Technology and Mathematics Academic Group in the Faculty of Education. He is the editor of the journal *Chemistry Education Research and Practice* (published by the Royal Society of Chemistry, RSC). He is on editorial, advisory and review boards for a range of journals, and he is the book reviews editor for *Studies in Science Education*. He was the RSC Teacher Fellow for 2000–2001. He has written books on aspects of science education and educational research, including *Chemical Misconceptions – prevention, diagnosis and cure* (2002); *Science: Teaching School Subjects 11–19* (2005, with Vanessa Kind); *Science Education for Gifted Learners* (2007, editor); *Enriching School Science for the Gifted Learner* (2007); *Classroom-based Research and Evidence-based Practice* (2007); *Progressing Science Education* (2009); and *Teaching Secondary Chemistry* (2012, editor). He has authored well over a hundred journal papers, articles for practitioner periodicals, and book chapters. Dr Taber taught science (mainly physics and chemistry) in secondary schools and further education. He undertook part-time study whilst teaching full time, preparing a master's dissertation on girls' under-representation in physics (a case study of a secondary school), and a doctoral thesis on A-level students' understanding of the chemical bond concept. He joined the Faculty of Education at Cambridge in 1999, initially to work primarily in teacher education. His current teaching is almost exclusively on higher degree courses, where he teaches educational research methods and supervises research students. His research is into aspects of student learning and thinking in science – including understanding of science concepts and aspects of the nature of science, and conceptual development and integration in science. Recently he has been involved in projects looking at lower secondary school teaching of physics topics; students' understanding of the relationship between science and religion, and young children's intuitive thinking about chemical phenomena.

Neil Taylor is Professor of Science and Technology Education at the University of New England, New South Wales, Australia. After completing a zoology degree in Belfast, he began his career in teaching as a science and geography teacher in Jamaica where he spent two years. He returned to the UK to undertake further study before taking up a position as a science teacher at a comprehensive school in Leicester. After four years in that position he moved to the University of the South Pacific (USP) where he conducted outreach work in the Pacific region on behalf of that university. In 1995 he left USP to complete a PhD at Queensland University of Technology researching primary teacher education in Fiji. He then had periods back at USP and later at the University of Leicester in the UK, before moving to his current university in Australia. His research interests include science and environmental education in developing countries, and he continues to assist with curriculum and resource development in the Pacific region, most recently in Fiji, Tonga and the Solomon Islands.

Rob Toplis taught science in inner-city, rural and suburban secondary schools for over twenty-five years. He is a Senior Lecturer at Brunel University, London, where he teaches and supervises on doctoral, masters and pre-service teacher programmes. His research and publications examine teaching and learning science in secondary schools with a particular emphasis on science for engagement and science for all. These interests include the science curriculum and curriculum change, pupils' attitudes, teaching scientific inquiry, and conceptual understanding and representation in chemistry education. He has emerging interests in the use of digital technologies and how they can relate to pedagogy.

Mike Watts is Professor of Education at Brunel University. He is active in writing and research in learning and teaching in higher education, and has long-standing research projects in this field. He was awarded a Higher Education Academy Fellowship in 2003 for his excellence in teaching and, in 2004, was elected a Fellow of the Institute of Physics. He is an external examiner for the National Universities of Ireland, and consultant to the Teaching Council of Ireland, as well as many other international duties. He enjoys 'naturalistic' classroom-based research projects in both formal and informal educational settings, evaluating the effects of policy and curriculum innovations on learning outcomes. Mike has carried out major studies of classroom interactions often, but not always, concerning the learning of science. His recent work has looked at how learners' own questions can be used as a basis for inquiry-based teaching, the ways in which feelings and emotions shape learning, and approaches to classroom technologies in learning processes. He has published widely and is keen to teach across all programmes at undergraduate, masters and doctoral level within education, and supervises a large number of PhD students. Being an avid, mildly rabid, devotee of Welsh rugby, his all-consuming interest leaves no spare time at all!

Acknowledgements

I am most grateful to Sue Capel for the opportunity to edit this book and, in doing so, slide in a few ideas of my own. Thank you, Sue, just the ticket. It is a privilege to work with friends old and new, and I thank all those who have contributed their thoughts and ideas here – and taken up the spirit of debate so willingly. I am particularly indebted to colleagues at Brunel for their conversations and camaraderie, and allowing our discussions to flow.

I would be utterly remiss not to thank my wonderful family, Siân, Andy, Rhian, Brian, Oscar, Rosie, Lily, Joel, Connor, Dylan, and to dedicate this book to my darling wife, Ruth.

Introduction to the series

This book, *Debates in Science Education*, is one of a series of books entitled *Debates in Subject Teaching*. The series has been designed to engage with a wide range of debates related to subject teaching. Unquestionably, debates vary among the subjects, but may include, for example, issues that:

- impact on Initial Teacher Education in the subject;
- are addressed in the classroom through the teaching of the subject;
- are related to the content of the subject and its definition;
- are related to subject pedagogy;
- are connected with the relationship between the subject and broader educational aims and objectives in society, and the philosophy and sociology of education;
- are related to the development of the subject and its future in the twenty-first century.

Consequently, each book presents key debates that subject teachers should understand, reflect on and engage in as part of their professional development. Chapters have been designed to highlight major questions, and to consider the evidence from research and practice in order to find possible answers. Some subject books or chapters offer at least one solution or a view of the ways forward, whereas others provide alternative views and leave readers to identify their own solution or view of the ways forward. The editors expect readers will want to pursue the issues raised, and so chapters include questions for further debate and suggestions for further reading. Debates covered in the series will provide the basis for discussion in university subject seminars or as topics for assignments or classroom research. The books have been written for all those with a professional interest in their subject, and, in particular: student teachers learning to teach the subject in secondary or primary school; newly qualified teachers; teachers undertaking study at Masters level; teachers with a subject coordination or leadership role, and those preparing for such responsibility; as well as mentors, university tutors, CPD organisers and advisers of the aforementioned groups.

Books in the series have a cross-phase dimension, because the editors believe that it is important for teachers in the primary, secondary and post-16 phases to look at subject teaching holistically, particularly in order to provide for continuity and progression, but also to increase their understanding of how children and young people learn. The balance of chapters that have a cross-phase relevance varies according to the issues relevant to different subjects. However, no matter where the emphasis is, the authors have drawn out the relevance of their topic to the whole of each book's intended audience.

Because of the range of the series, both in terms of the issues covered and its cross-phase concern, each book is an edited collection. Editors have commissioned new writing from experts on particular issues, who, collectively, represent many different perspectives on subject teaching. Readers should not expect a book in this series to cover the entire range of debates relevant to the subject, or to offer a completely unified view of subject teaching, or that every debate will be dealt with discretely, or that all aspects of a debate will be covered. Part of what each book in this series offers to readers is the opportunity to explore the interrelationships between positions in debates and, indeed, among the debates themselves, by identifying the overlapping concerns and competing arguments that are woven through the text.

The editors are aware that many initiatives in subject teaching continue to originate from the centre, and that teachers have decreasing control of subject content, pedagogy and assessment strategies. The editors strongly believe that for teaching to remain properly a vocation and a profession, teachers must be invited to be part of a creative and critical dialogue about subject teaching, and should be encouraged to reflect, criticise, problem-solve and innovate. This series is intended to provide teachers with a stimulus for democratic involvement in the development of the discourse of subject teaching.

Susan Capel, Jon Davison, James Arthur and John Moss
December 2010

Part I

Introduction

Chapter 1

Introducing the debates

Mike Watts

I was in full swing, cooking a summer barbecue with the indispensable help of Lily, my beautiful five-year-old granddaughter. She was chatting gaily while I flipped burgers, sausages, chicken and, at one point, I bent down to pay her rapt attention, full of lofty grandfatherly wisdom. As she looked up at me, earnest big brown eyes, she asked, 'Taid, why do you have hairs up your nose?'

Taid is Welsh for granddad, but quite why I have copious nose hair is far less easy to translate. A little later, eating said burgers and sausage, one of the family asked just what was the point of this book I was editing. I began by talking, with a fair degree of hand-waving, on the nature of science and technology, the role of education, of constructivism and conceptual challenge – and this stirred a cross-table debate about science's certainties, its fixed and factual methods and content, the trust we place in, say, medical science, genetics or engineering. This was countered round the table by the sceptics, the artists, the faithful and the folklorists. Science is powerful, potent and essential; science is esoteric, expensive and irrelevant. Science is reliable knowledge and truth; science is western, white and male. It's a naturally argumentative family!

I was not neutral in the conversation. My first love and affiliation is physics, though my devotion is not blind and uncritical. In the heat of the exchanges, arms gesticulating, I became ever more convinced that debate lies at the very heart of science. I manoeuvred my way through the flurry of animated issues while coping with a char-grilled rib, and I hung on to the notion that science is the project of discovering and generating the fundamental ideas we need for life, and about life. For me, it is an intimate project: I want to make sense of the science that others both love and loath; I want to understand its freedoms and limitations, its philosophy and powerful relevance to everyday living. Oh, and I do enjoy the debate. For me, it is invaluable: sometimes it is actually easier to be the questioner, to investigate what other people think than to work out my own position; it is harder to judge the personal meaning and value of an issue than it is to record the variations in what other people say and do.

But that said, my second love is education. I am a learner and a teacher bred in the bone. Even while I debated round the table hot issues in health and diets, genome projects and Large Hadron Colliders, navigated between science fact and science fiction, flicked crumbs from my jumper, wiped grease and marinade off my fingers, I was actually distracted. I wanted to pull Lily away from playing with the other children and explain to her exactly why we all have hair up our noses. In part I wanted to make up for my stumbling ineptitude midst the cooking, in part because she'd asked an intriguing question and, in part, because I didn't actually know the answer. I trusted to my long-held belief that a certain way to reach my own understanding is to try and explain it to someone else. In common with many teachers, I often don't know quite what I think until I hear myself say it. Telling other people is a prelude to discovering my own ideas. Are the hairs a first filter for airborne particles? Why the evolutionary need? Our lungs are fairly robust: mucus collects and copes with most invasive dirt and dust. Noses are not just for breathing but we presumably need an alternative air-intake system given the complex jobs our mouths are asked to do (eating, drinking, initial digestion, tasting, communicating, kissing etc.). So why the hairs in the nose? We don't have hairs in our mouths. Why more nose hair as we get older? Children have so little – Why, at an age when they may need it most? Why do males have more than females? Good question, Lily, now just come here a moment, be quiet and be utterly enthralled while I lecture you to enlighten myself.

What is debate? At its core, debate is an extended argument. Two (or more) people, or groups of people dispute whether a statement, situation, state of affairs is true or false, right or wrong. On the whole, one group supports the statement, while the other opposes. In a formal debate the participants and the audience finally cast a vote and the matter is then laid to rest, at least for a while. Parliamentary debates, even televised political debates between presidential candidates, move inexorably to democratic dénouement. Not so in our family. And seldom so in science education.

Science and education, on opposite sides of the fence? The first, an appreciation of the grandeur and depth of symbols with only a passing and cautious regard for the contribution of individual intellect. The second, grounded in the intense awareness of individual consciousness and practice, and a cautious approach to symbolic systems. The nature of a free debate is just that, the freedom to think and say what you like so long as there is no infringement of the rights of others. It is the freedom to work out one's own understandings, position, to decide what you believe and why, to create and follow your own vision of science and education. We are entitled to our own views, be they illogical, anecdotal, magical, fanciful or purely mythical. This is a profound liberty based securely in the freedom to realise that truths we discover for ourselves are hugely more valuable than those we merely accept on external authority. We think for ourselves, no one can tell us what to think, or impose their will on us. Well, except for teachers, maybe. But the best of teachers do not simply replace one set of stock notions with others but challenge learners' thinking to arrive at new notions for themselves. They provide the mental space for stimulation, hesitation, reflection, as well as guidance, comparison and wonder. As a teacher I am commonly taken up by trying to reform the world: why on earth (or elsewhere?) do so few students see the pristine beauty and poetry of physics?

The field of education is highly contested terrain. Discourse within education is at the same time both grounded in everyday life with real and lively young people with young and lively minds, as well as being sometimes opaque and esoteric to the non-specialist. This can particularly be the case if the subject is discussed in ways that are non-interactive and disengaged. The chapters in this book are designed quite specifically to engage the reader in debate, an invitation to be challenged, to decide on a perspective, read the opposing viewpoints and expect to be challenged all over again. Keith Taber sets the ball rolling by looking at the audience for science education, who are they and what exactly is wanted of them. What does the government need? What does the slogan 'science for all' actually mean? And just what kind of science are we talking about? Chris Lloyd-Staples picks up this theme from a more local perspective, and opens with a taxing set of questions: whether changes in science education can be driven by short-term government policy initiatives. Or, whether deeper changes require a sustained focus on an objective for maybe a decade or more, in order to overcome the inherent (?) resistance of schools and teachers. Given that most governments have a maximum of five years of electoral mandate at any one time, few of them – local or national – can live easily with processes of educational change that are decades long. That said, on the other side, what's worth preserving is worth preserving, and rapid change seldom embeds well. Are teachers, science teachers, naturally resistant to change? It would seem so.

Joanne Cole discusses one key part of the audience and asks why, after all these years and considerable effort, girls are still in a minority in science. But, what if the reason for the under-representation of women in STEM (science, technology, engineering and mathematics) is simply that girls are not as good at science as boys? That's possible, surely? We know that girls lack the essentials to become physicists. Don't we? Having asked these questions she then goes on to answer them and suggest ways in which it might be possible to reverse the under-representation of young women in science. Michael Allen raises another perennial problem, when children transfer from primary to secondary school 'something appears to be lost … that is never regained'. He discusses a range of reasons why young pupils can be very disappointed with their initiation into secondary science, and why it does not live up to what they expected. This leads them to switch off from science and, as their attitudes deteriorate, achievement levels can fall. He then sets out some approaches that might help to reverse this dip in initial interest and attainment. Can we possibly mend 'the dip'? As Chris Lloyd-Staples has already pointed out, it is not always possible to look to government for the solutions. Clear food for thought.

In Chapter 6, Rob Toplis picks up the 'curriculum story' and discusses how science works, both in principle and in classroom practice. In particular he looks at the four main sections of the *How Science Works* strand of the National Curriculum for England, and discusses teachers' and pupils' perceptions of what is entailed.

In the next part of the book we move closer to the classroom. Nic Crowe comes at things from two very different perspectives: from the pupil's point of view, and from their interest in online gaming. His extracts show that young people's talk is frank and to the point – and leave a powerful question: what exactly does it take to engage young people in science? Surely, crude video games are the very antithesis of structured rational science? Susan Rodrigues tackles the issue of language in science classrooms: what can be more alienating than having a teacher who talks in a foreign tongue? One who talks incomprehensible chemistry! Clearly there is need to come to terms (quite literally) with a technical language (isn't there?), but what ever happened to plain English?

Helena Pedrosa-de-Jesus and Sara Leite show how they create engagement in the context of chemistry lessons, by pulling in pupils through their questions, and making time for their search for explanations and understanding. I follow this with a further look at some of the 'ins and outs' of explanations, and why school science needs explanations at all. It does, doesn't it? Or is that simply a bridge too far – that students should actually *understand* the science they are doing? Does not understanding come so much later, once you've got the basics? Around the time when you start a PhD? The school assessment system is simply not geared to measure understanding, is it? Then John Gilbert focuses on the nature of context, what it means and why we chase it. His questions are direct and to the point: context means different things to different teachers and curriculum-makers; it is difficult to achieve, so why bother?

Richard Malthouse and Jodi Roffey-Barentsen discuss the tension between science and reflective practice. It is a tension that can occur when a scientist first trains to become a teacher, or subsequently engages in professional development. Is s(he) actually a scientist or a teacher? There may be a conflict of purpose between doing science and 'doing' science education – in which reflective practice is an inherent part. Richard and Jodi offer a new model of reflective practice that they call Situated Reflective Practice (SRP), and set this alongside a range of strategies that assist reflective practice in groups. Can science teachers reflect? It would appear not.

David Barlex and Marion Rutland take the issues into broader territory: what is the relationship between science and technology? Does science simply provide the 'service-need' for technology, or is it the other way round – technology is simply the 'appliance of science'? They illustrate their perspective through some stirring examples. And then Neil Taylor, Frances Quinn and Judith Miller pick up the thorniest of issues: exactly who should teach sex education? And how? This is an area not short on controversy, and a debate not confined to any one country, as they very clearly point out. But if Neil and colleagues place the body centre-stage in science, then Steve Alsop asks if it really is there. This is a wonderfully philosophical piece, yet securely rooted in the very ways in which we teach 'disembodied' science. Finally Peeranut Kanhadilok puts science firmly in its place: it is just one of many knowledge systems and must take its turn alongside other systems. Nor is it necessary to integrate science with local wisdom: the role of science education is to be there when needed, alongside other cultural and social systems.

The science of today cannot be separated from its application in society, and I am always on constant look-out for ways in which to embed school science, make it contemporary, make it relevant. I am not a great sailor, but four of us (family and friends) once spent a few days yachting in the Mediterranean near Elba off the coast of Italy. When physics inevitably reared its beautiful head, we fell to pondering whether it was actually possible to 'heave to', come to a complete and utter stop in the middle of the water, regardless of current, wind, tide or steering. The nautical practice is to set the sails against the keel and the rudder, so the forces on the boat fully cancel each other. Once stopped, there is no resultant force and so there is then no movement at all. Relative to what? Is it actually possible? We were out of sight of land. Did it work? Does the theory of 'boating physics' truly come real in practical situations? Surely there would be some drift, *some* movement? Ed (captain by simple dint of owning the boat) organised the sailing mechanics and we tested the principles over an hour while we calmly sat on deck in the warm sunshine, ate fried Irish sausages and drank hot Bovril. Yes, relative to Elba, Italy and the rest of planet (and elsewhere?), the boat stayed very still in position, was wholly unmoving as testified by the boat's (infallible?) GPS system. Physics is truly poetic, *non è vero*? The sausages and Bovril were good, too.

Part II

Policy debates

Chapter 2

'Science for all' or science education for the good of all?

Keith S. Taber

Introduction

For several decades 'science for all' has been something of a clarion call in science education – suggesting that science is a subject suitable and necessary for all learners, which should be a core and sacrosanct part of the curriculum. Yet we also know that many students consider science too difficult, or find it boring, especially when that science is taught as something abstract, theoretical and mathematicalised. This is sometimes countered by finding ways to make science relevant: by teaching it in contexts that are assumed to be motivating; by using concrete examples to make the ideas less abstract; by avoiding too much theory; and – wherever possible – avoiding mathematics.

That is all very well if making school science accessible to all is more important than enthusing and challenging those capable of high achievement in science. Yet, by its nature, science is about abstraction; it is inherently theoretical, and much (although not all) of it is best represented in mathematical forms. So it could be argued that a science education that is concrete, a-theoretical and purely qualitative is not an authentic science education at all. Moreover, it can be suggested, such a bowdlerised science education is only accessible at the cost of being totally unfit-for-purpose for those students that might have the aptitude to study 'proper science' and so might potentially go on to become the scientists, engineers and technologists we need for a sustainable and comfortable future.

Indeed, those gifted students who have the greatest potential for high achievement in science will find such a 'theory-lite' version of science undemanding and intellectually unsatisfying, and so may look for academic challenge in other areas of the curriculum able to offer them genuine intellectual engagement (Taber 2007b). So is 'science for all' just an ideological sop which offers the pretence of providing everyone with a scientific education, whilst actually undermining a *genuine* science education for those individuals we most need to be attracted into science? This is the core question at the heart of the debate explored in this chapter.

The themes of this chapter will be illustrated by a reflection upon some of the changes seen in the English education system over the past half-century. This provides an excellent case for consideration – as it has been (and remains) subject to many opposing influences, and incongruous directives from government, that have led to almost constant (and seldom completed) shifts in direction. The account presented here is intended to be an honest simplification of a rather complex topic. However, even though it is designed to be honest, it is inevitably coloured by the values and experiences of the author. My intention is not to suggest there is a clear solution to the key issues considered here (especially related to when and how it might be both educationally effective *and* fair to divide students by attainment or perceived ability in schools), but to try and offer some background to the argument, especially in relation to school science teaching.

So what is the problem?

The crux of the problem motivating this debate is that there are actually two related but quite different *underlying* problems that governments are seeking to address simultaneously. As these are rather different problems, it is not clear that the same solution will work for both. Both problems relate to perceived deficiencies.

Problem 1: The population is not 'scientifically literate'

There has long been an argument that too many people do not understand the basics of science, and indeed that it has often been seen as perfectly acceptable for someone to be considered an educated and cultured person without knowing any science (Snow 1959/1998). Lack of familiarity with Shakespeare or Mozart is considered a sign of being uncultured, whereas – the common perception suggested – something as esoteric as thermodynamics is only of interest to a few rather sad people lacking social skills, and having rather particular interests. In the United States, concern about public understanding of science has been especially acute, when surveys regularly show that large sections of the population doubt that people could have walked on the moon, but are quite happy to believe that people once walked with dinosaurs.

Problem 2: There are not enough talented people being attracted into advanced study and careers in science-based fields

Governments and learned societies have been concerned about the numbers of able and well-prepared candidates applying for university study in many areas of science for some time. In the 1990s, competition for the available UK university places on physics degree courses became so limited that candidates with two modest passes at A level (the most common entrance qualification) were accepted in some institutions, and it became common to offer top-up courses before or during the first undergraduate year to provide the background that would traditionally have been assumed on entry. Departments of physics and chemistry were closed in some of the more prestigious universities, suggesting these subjects were no longer seen as disciplines that all good universities should include in the curriculum.

So governments are worried both that people *in general* do not understand science well enough; and that not enough *particular* people who might make good scientists are choosing to follow that path. Both of these problems are seen as important because it is considered they reflect potential impediments to providing safe and comfortable lives for people in the future. This argument is based on several widely accepted claims:

- The current standard of living of people in the so-called developed world depends upon a high level of resource (power supply, materials) that is unsustainable in the long term with current technology and means of production.
- Current levels of human activity have caused extensive environmental damage through pollution, and loss of habitat (and so reduced biodiversity), as well as ongoing climatic change that may alter weather patterns, crop yields, sea level etc.
- The so called 'developing' nations aspire to similar consumption as the richer countries, and so the global per capita use of energy supplies and other resources is likely to increase substantially as economic development of these countries proceeds.
- Increasing levels of production that are based on current technologies will inevitably only be possible with increased damage to local and the global environment.

Clearly a lot more could be said about each of these propositions, but readers will probably recognise that these are robust assumptions and rightly major areas for concern.

Education to the rescue?

Now if these are seen as the fundamental problems, then science and education are considered important elements of possible solutions. Indeed there are precedents here. It is well established, for example, that increased levels of education tend to change demographics: better educated populations tend on average to produce fewer (but better educated) children. So education may help to at least avoid population 'explosions' that drive more production and consumption of goods as countries develop. This might offset to some extent the inevitable increase in per capita resource consumption as such nations become wealthier.

Science education is seen as especially important because – the argument goes – people need to be enabled to make informed personal decisions (about what they choose to buy, how they treat the environment, how they dispose of rubbish etc.), and civic decisions (about voting choices for example) that take into account an understanding of the available scientific evidence. So, it might be mooted that worries about the possible risks from nuclear power stations are irrational when compared with the likely risks to society of failing to build sufficient power-generation capacity to keep the lights on as accessible supplies of fossil fuels become exhausted. (It may well be that reserves of such resources have been underestimated, and new technologies may make currently inaccessible or uneconomic fields viable in future, but that would only delay the inevitable crisis.) That is clearly not a straightforward matter: whilst people have been killed and harmed in 'nuclear accidents', the numbers of direct casualties are small compared with fatalities involved in such activities as coalmining. But obviously these are nuanced issues: people being asked to vote in informed ways for policies need to be informed about and understand the risks (to people and the environment) involved in mining and transport of materials, the storage of waste, the potential for terrorists acquiring material, and possible nuclear weapons proliferation, as well as the safety of nuclear plants themselves. This is without considering economic issues relating to construction and eventual decommissioning of the plant.

Science to the rescue?

The second theme relates to how we manage to provide people with the levels of consumption (and so, it is argued, 'quality of life') we are familiar with in the richer nations, without running out of resources and ruining the environment – and here science is expected to provide new ideas, and the technologies they can facilitate. Up to now, science has often come to the rescue. Nineteenth-century predictions that the streets of London and other major cities would soon be submerged in horse excrement due to increasing levels of traffic proved to be unduly pessimistic; instead London streets are full of slow moving motor vehicles, burning up fossils fuels and producing both engine noise and the substitute excrement of poisons such as nitrogen oxides, carbon monoxide and ozone. Whether horse manure, which could be recycled, was preferable is a moot point.

The work of research scientists makes possible new materials and more efficient technologies, and many of these are much 'greener' than the options they can replace (Emsley 2010). The rate of technological progress experienced by people living today is quite unlike any previous period of human history. Wind farms, MP3 players, mobile phones, the Internet, self-cleaning windows, low-power bright lamps, disease resistant crops, improved drugs with greater specificity and fewer side effects, clothes that do not need ironing ... the list of technological products based on the work of scientists is very long indeed. Some make life more fun or convenient – some may protect or extend life itself. Of course, science also feeds into technologies that can, for example, locate buildings and kill their occupants from a great distance – as always, the application of science offers choices that are informed by priorities and values that are outside science itself.

There is little doubt that science is contributing to solutions to major problems by producing: new 'greener' materials (e.g. from plant biomass) that do not rely upon scarce resources; renewable forms of power supply; and new forms of production that are more energy efficient and produce less toxic waste. Scientists are also working hard on issues such as understanding climate change, protecting habitats and endangered species, and finding out how to protect people against epidemic diseases, crop failures, earthquakes, rising sea levels and so on.

But rescue attempts may be thwarted

There is however a significant doubt about whether the work of science is proceeding fast enough to counter the problems the world's population faces – or whether ultimately science can at best only ease the extent of these major problems. We have not yet solved the 'energy crisis' that was first a major cause of public concern in the 1970s – although progress has certainly been made. With such major issues it is not always clear whether the limited degree of progress reflects the inherent limitations on the pace of science itself, the lack of political will among governments, or the failure to persuade electorates to pressure their politicians to prioritise what are clearly very serious problems. Socio-scientific issues are inherently complex because, even when the science is well understood, decisions about how to act on scientific knowledge are influenced by a great many extra-scientific factors. Public ignorance can certainly be a concern, but the immediate self-interests of individuals and pressure groups (such as the wealthy who may not wish to commit more income to taxation; such as multinational companies who profit from current ways of doing things) are also pertinent.

Scientific knowledge is sometimes thought to be value-free – but its application certainly depends upon value judgements. If wind farms are seen as ugly (a perfectly admissible aesthetic consideration, although not one this author shares) and unnatural 'blots on the landscape', it becomes difficult to persuade people to accept them in their locale. It is over forty years since it became recognised that tobacco is both an addictive material that most people find extremely difficult to 'kick' once they are regular smokers, and one of the major sources of avoidable serious illness and early death. Yet despite smoking becoming gradually more marginalised and unacceptable in many developed nations, the large tobacco companies continue to profit by peddling the drug, aiming much of their marketing at developing, less regulated, national economies. The scientific evidence is quite clear, and if tobacco was discovered today, then those seeking to profit by importing the material to sell to addicts would be hunted by the police and other government agencies with the same fervour used to destroy poppy harvests in Afghanistan and the activities of drug cartels in Latin America. Yet political expediency allows tobacco companies to continue to profit from the international trade in an addictive and harmful drug. There are many other examples where scientific evidence is available, but the scientific argument does not win the day. At the time of writing, the British government has just approved a cull of badgers to help control the spread of bovine tuberculosis, despite clear scientific advice that this will not be effective. This is an example of how, even when governments ask for scientific advice, they may not have the courage to follow it, because of the influence of pressure groups who feel they know better than the scientists.

The recruitment imperative

Regardless of the extent to which the application of scientific advances is moderated by extra-scientific considerations, it is obvious that *continued* scientific progress depends upon talented young people choosing to study science subjects at the end of compulsory schooling, and then deciding to use their scientific education to follow scientific careers. Quite rightly, governments are concerned that the education system should produce enough people who have the potential, and the interest, to want to become scientists, engineers and technologists. It is imperative that enough of these people take qualifying examinations and seek higher education in science-based disciplines to provide a supply of keen and capable science undergraduates. It is then also imperative that enough of these students do well in their scientific education and wish to develop a career in scientific research. And there also have to be enough other strong, keen science graduates who wish to enter science teaching, or work in other areas where it is important there are science specialists: the media, the law the civil service and so forth.

The traditional strategy for ensuring a supply of scientific personnel

The education system that developed in England in the middle of the twentieth century could be considered to have been well suited to ensuring a supply of suitable candidates for scientific professions. It had a number of features that might be considered to have worked together to ensure that schooling provided a flow of capable young people entering university to study the sciences. Anyone who knows about the English education system will be aware it is a rather complex and messy system, so that any general statements are necessarily simplifications, but the following points can be considered to be generally accurate:

- There were several types of schools intended for students with different aptitudes. Grammar schools selected the most 'academic' students. Usually children were tested at age 11 (the 'eleven-plus' examination). Those judged most 'intelligent' went to the grammar schools, where they mixed with other 'intelligent' children, and were taught by teachers who were experienced in working with 'academic' children through pedagogy that suited 'academic' learners. Often within the schools, there were streams so that the most able could work with others of similar ability, allowing them to stimulate each other, and allowing teachers to teach in depth, and at a pace suited to more 'gifted' students. The schools prepared students for academic examinations in a range of different subjects: examinations that in the case of science subjects had a strong emphasis on scientific knowledge and understanding.

- There was no National Curriculum, so schools were able to allow students some choice of study subjects. Students who were interested in the sciences could follow separate examination courses in biology, chemistry and physics (and sometimes additional science subjects such as astronomy or geology). Teachers generally specialised in their own subject – so, for example, biology specialists were not expected to prepare students for the physics examinations. At age 16 (when many non-academic students had already left the school system), those who were able and interested in science could demonstrate their learning in examinations which were designed so that they could only be passed by able, studious learners, and which awarded a range of grades to distinguish different levels of achievement even among this successful group.

- The grammar schools had 'sixth forms' where students prepared for the further, more advanced examinations for university entrance. Because of the academic nature of the schools, sufficient numbers of students 'stayed on' to make viable groups across a range of academic subjects. There was often a sense of sixth forms having separate science and humanities routes, so that there was a clearly-identifiably cohort of 'scientists' studying together (Hudson 1967). The examinations at the end of sixth form were again designed to be academically challenging, on the assumption that the examinations previously taken at age 16 had already acted as a filter to allow only those suitable for advanced study to continue.

It has been argued that the academic school curriculum in grammar schools was to a considerable degree informed by the needs of the universities. So the university entrance examinations were designed to see who was prepared for undergraduate study, the sixth-form curriculum being designed accordingly. Similarly, the examinations used to select students for sixth-form studies were designed to identify those who could demonstrate knowledge and understanding of school subjects at a sufficient level to be ready to make further progress. The grammar school curriculum could then be designed (by the school teachers) to prepare the students entering the schools at age 11 (based on a competitive examination) for the academically-orientated examinations they would take five years later. Primary schools would see part of their role as preparing students for the examination taken at age 11 that would select which students were suitable to enter this academic stream that could often take them to sixth form, and beyond to university study (and so, potentially, to professional careers).

Other students studied in different ('modern' and sometimes 'technical') schools, with a more diverse curriculum, including an emphasis on craft and applied subjects (e.g. rural science, automotive engineering), leading to examinations that recognised a much broader range of achievement, and allowed students judged to be of more limited intellectual ability to achieve some success. Teachers in these schools were under less pressure to focus on the difficult abstract concepts that would be important for passing examinations at the end of grammar school, allowing for flexibility in topics (especially where examinations were developed locally, through what was referred to as 'mode 3') and more project and practical work.

The winds of change

In many ways, the grammar school system worked well ... at least for those students who were 'academically minded', and for those teachers who enjoyed working with students who would tend to challenge them, not through disengagement or misbehaviour but through probing questions about subject matter. The system also ensured that universities were supplied with students who were well prepared (at least in terms of subject matter) to proceed on to undergraduate work in both science and humanities disciplines. The strengths of this system have led to the current Secretary of State (i.e. the senior cabinet minister) for Education in England to aspire to reinstate something of that regime as a solution to perceived failures of the current education system (Department for Education 2011).

Over several decades the English system saw significant shifts that led to secondary schools that were generally not able to select students on grounds of ability, a common curriculum system that requires all students to take the same core subjects, leading to examinations intended to recognise both high intellectual achievement of the few, and yet also accredit the more limited achievements of the rest of the population. Part of that core curriculum has been 'science', a subject that became compulsory for all students of ages 5–16, and for which there was much the same curriculum for all learners within each 'Key Stage'. Associated with this was an assumption that if 'science' was the school subject, then teacher preparation needed to be for 'science' teachers (rather than chemistry teachers, physics teachers etc.).

So there was a swing from a very differentiated and hierarchical education system that attempted to identify different 'types' of students early, to channel them into specialised provision which, for those considered more academic, was based on preparation to study recognised academic subjects at university, to a system where virtually all children were expected to be taught and then examined on the same spread of science topics, without regard to ability and without allowing a choice to specialise (or drop) particular science subjects.

There were very good reasons to abandon the divisive examination at age 11 which – based on one morning's work – could allow a child access to an academic education, or direct them to a different flavour of schooling that usually made it very hard for them to ever switch back to the academic track (Taber 2012). In particular, it was recognised that levels of success on such examinations often reflected the degree of advantage provided by upbringing more than intellectual potential. If bringing back grammar schools means once again asking children to take competitive examinations with such high stakes at such a young age, then this would indeed be a retrograde step.

It is important to recognise that ideology has an important part to play in issues such as this. If you believe that 'breeding will out' and that children from 'middle-class' backgrounds probably carry genes 'for' academic success, then a return to a bipartite system of education (subject to the proviso of a safety net for the occasional genius who gets exam nerves or has flu on exam day) might seem attractive. However, if you believe that most human beings have the potential to develop strong intellectual abilities given the right support, but that the affordances of different home environments can mask true academic potential, then you might feel that any decisions that classify people in apparently permanent ways are not only ill-advised (and against the interests of the common good), but fundamentally wrong from a moral perspective.

Modern science tells us that to think of genes for academic success is inappropriate: complex traits like intelligence – which itself is a contended notion (Gould 1992) – result from the interactions between a range of genetic factors and the environment in complex ways, such that there is little sense in ascribing positive or negative qualities to individual genes in isolation. Despite this, the extent to which being considered intelligent or academic or gifted or a genius etc. might be constructed as something that is in part inherited remains a live debate where there is no simple scientific consensus. Inevitably, what evidence there is will be coloured by the values and cultural norms that individuals bring to their interpretations.

A modern day Thomas Gray (the poet who mused on what might have been achieved by the illiterate farmhands buried in a country churchyard, had they been given different opportunities) might write a telling educational elegy on those passing out of the second(-class)ary modern school, marked as most suited to be shop assistants or telesales operatives.

> Perhaps in this neglected playground was spent
> Some genius hidden in secondary modern guise
> Hands that might have written a sublime sonnet
> Or earned that mind a Nobel prize

Yet some might feel the potential waste of talent of those who could do well with encouragement and support – but who are *not* identified as academically able at the end of primary school – a price worth paying if it allows a rigorous, intellectually-demanding science education for enough able students to provide for the future needs of society. After all, those of this mind may argue, the needs of the many may require the sacrifice of a few. (Perhaps especially if I am secure that my children have been provided with all the advantages I can afford to ensure they are among the select.)

To set, or not to set?

So, in the 1970s, most schools in England shifted to being 'comprehensive': accepting students of all abilities from their local community. These schools adopted the two different examination systems that had originally been intended for grammar school and secondary modern school students. Many of these 'comps' had sixth forms, and these would often include both the traditional academic subjects and courses more suitable for other students who chose to stay on at school. As part of the comprehensive movement, there was a strong drive towards other 'progressive' approaches that fitted the comprehensive nature of the school populations. So 'mixed-ability' teaching was popular, allowing students of very different levels of attainment to study together in the same class (White 1987).

However, mixed-ability teaching was not always applied in all subjects or to all year groups. After all, as long as all students were in the same school, and there was the possibility to move between classes, some 'setting' need not permanently exclude anyone from opportunities to demonstrate academic potential. In principle at least, students who might be 'late developers' (or decide to increase their level of engagement with school work) could be moved between sets where this seemed appropriate. Readers will probably recognise potential complications here though. Transfer between sets was only likely to be effective if the different sets were studying the same material, at much the same level – which of course would make the setting rather pointless. So, students to be 'promoted' needed to be seen to be capable of both (1) the higher-level work and (2) doing any needed 'catching-up'. A second issue concerns the effects of labelling, in that being assigned to a set could inform students' own expectations of what they might be capable off, as well as indicate what they were expected to achieve.

Arguments about the relative strengths of setting and mixed-ability teaching continue: because mixed-ability teaching can mean many things; because making fair comparisons between different educational conditions is difficult (Taber 2013); and because different values identify different possible educational aims as important. (For example, are social integration and cohesion, learning toleration for others different to ourselves, and understanding diversity, more *or* less important than optimising examination results?)

It also seems that concerns with the needs of gifted children, something that had been an issue in at least some schools in the 1960s, tended to go out of fashion with the comprehensive regime. Primary schools no longer had to worry about potentially disadvantaged students with latent ability that might not be revealed by the eleven-plus examination, and who might miss out on grammar school. In secondary schools, the ideology of mixed-ability teaching suggested that identifying some students for special treatment was unnecessary and inappropriate. Perhaps it was commonly thought that assigning scarce resources to give further advantage to those who were already seen to be gifted was less a priority than using available resources to support the learning of those with special needs that would be an impediment to keeping up in class. After all, the talented teachers would be differentiating lessons in creative ways, and any less creative mixed-ability teaching that offered the same fare for all should not be a problem for such students. They could be challenged by helping their less fortunate classmates – and if they were really gifted they would surely be able find their own ways to progress their learning without direction from the teacher (Taber 2009a).

Finding a science for all

In the 1980s, the two separate public examinations taken at age 16 in England were replaced by a new examination intended to meet the original purposes of both – that is, to offer certification of a range of levels of attainment that both distinguished at the highest levels, and offered recognition of much lower levels of academic achievement. It soon became expected that virtually all students should be able to achieve some kind of pass in the new combined General Certificate of Secondary Education (GCSE).

A common examination meant that there needed to be a common curriculum that the full ability range could engage with, and common examination papers that allowed all students some success (whilst still distinguishing the highest achievers). However, at this time there was still considerable scope for students to select examination subjects in most schools. So although the GCSE Physics, for example, was meant to be suitable for virtually all students, many would opt out of physics at age 14 and select instead something more to their liking. However, pressure groups, such as the Association for Science Education, the professional organisation for science teachers in the United Kingdom, argued that all students ought to be taught a 'broad and balanced' science throughout their school career (Assocation for Science Education 1981), and this informed government policy that 'science for all pupils up to the age of 16 should include coverage at an appropriate level of the basic concepts of biology, chemistry and physics' (Department of Education and Science & Welsh Office 1985: 12).

At the end of the 1980s, the government introduced a National Curriculum (Statutory Instrument 1989). As part of this, 'science' became a compulsory subject for all 5–16 year-olds. The secondary curriculum was intended to cover what was seen as essential in biology, chemistry, physics and some earth science, as well as understanding something of the nature of science, and, as a result, included a good many different topics that *all* secondary students had to be taught. For most students, the expectation was that an examination course 'worth' two GCSEs would be taken: the 'GCSE double science' had been created. Although this could be 'topped-up' by those especially keen to take extra science examinations to convert it to 'triple science' (i.e. certification in biology, chemistry and physics), the argument that students should experience curriculum breadth limited the numbers taking triple science, and in many secondary schools this option was not offered.

Why 'science for all' did not work for all

There is nothing inherently wrong with the idea that science is for all pupils, but the notion that the same form of science education is suitable for all students proved problematic. Many 'less academic' students, and those with limited interest in science subjects, found much of the material – a constant carousel of many different science topics – difficult, boring and irrelevant (Cerini *et al.* 2003). However, the lack of depth (necessary to squeeze so many topics into a double subject) made the new subject intellectually undemanding for those with genuine interests and ability in the sciences (Osborne and Collins 2001). The need to provide examination questions that would allow all some sense of achievement tended to squeeze out the more difficult concepts, and in particular anything that required more than the most basic mathematics, as well as questions that required the candidate to plan and structure extended answers (as had been common in the grammar school examinations).

Moreover, attempts to get across the nature of the scientific processes, seen as a key element in developing 'scientific literacy' for informed citizens, was judged to have had limited success, despite several attempts to patch-up this aspect of the curriculum. Initial plans to include such a theme as a discrete part of the curriculum had been diluted and reduced to a focus on scientific investigation that – because of assessment requirements – soon became reduced in practice to teaching a trivial and unauthentic notion of fair testing (Taber 2008). It became clear that the curriculum – heavy on specification of a good deal of scientific knowledge – did not allow effective teaching about the nature of science itself.

Science only fit for the pub?

Recognition that the National Curriculum for science was not fit for purpose (Millar and Osborne 1998) eventually led to a drastic change of direction from a comprehensive specification of content (DfEE/QCA 1999) to a brief outline of key topics to be included, and a much greater emphasis on 'how science works' (QCA 2007a, 2007b). The adoption of new examination courses led to public perceptions that school science had become taken over by low-level discussion work on 'issues'. This led to a national debate about whether this latest version of 'science for all' was only fit for preparing future citizens for informal conversations about scientific topics over a drink in their local public house (Gilland 2006).

Teaching in interesting times

Most of the trends discussed above have been halted or reversed to some extent. The shift to comprehensive education and more mixed-ability teaching was moderated by setting within non-selective schools, and in recent years there have been repeated efforts to square the circle of allowing all schools to be in some sense selective and 'academically-oriented' whilst remaining potentially open to all. The single examination system soon admitted tiered examinations in some subjects, and an additional 'starred' grade was added to recognise the highest level of attainment. The uptake of triple science increased, especially after the government suggested it was an entitlement for those students identified as being able in science by age 14. At the time of writing, a(nother!) new curriculum is being planned, which is unlikely to bear much resemblance to the present version. Expectations are that it will look more like a grammar school curriculum – this might well meet the needs of the most able students, and help ensure they are engaged and motivated to read sciences at university.

Whether that must be at the cost of boring, confusing and failing most of the ability range in secondary schools remains to be seen – arguably that depends on whether the government is prepared to acknowledge the lessons from the past half-century of changes. If everyone is expected to study a highly-conceptual, abstract, theory-rich, mathematicalised (and so authentic and rigorous) science, then we will safeguard the future supply of scientists, at the risk of disengaging and possibly alienating most future citizens from science. They may not only fail to become science literate: they may actually reject science and become more readily persuaded by snake-oil sales people peddling, *inter alia*, crystal healing, young earth creationism, concepts of national bloodlines, fear of all radiation, and claims that global warming is just a conspiracy propagated by dangerous left-wing radicals (whilst the kinds of molecular radicals actually feeding photochemical reactions damaging our atmosphere are dismissed as scientific mumbo-jumbo).

What if there is no panacea?

Perhaps it just is not possible to have a science curriculum which challenges and motivates the gifted students while simultaneously allowing students of all abilities to understand and achieve – at least not if you want that curriculum to be uniform and open-to-all. Surely there has to be a compromise between how we ensure opportunity for all, and how we recognise the needs of different groups of learners? It does not make sense to *start* science topics with mathematical treatments (Taber 2009b), but there has to be potential for some learners to progress from the qualitative understanding of concepts to mathematical representation.

We are not going to prepare the future research scientists we need by turning science lessons into versions of those television debate shows, where all views are aired but no opinions are developed. Yet, even if a continuous diet of issue-based talking-shops is unlikely to be best preparation for scientists, it is certainly the case that future scientists will need to engage with socio-scientific issues and public debate. Perhaps 'how science works' did not work as an organising focus for secondary science teaching any more than the identification of theoretical keys ideas (e.g. energy, interdependence) had helped students understand the nature of science in the previous incarnation of the curriculum. Yet it would be a big mistake to drop the focus on the nature of science to return to a menu of discrete (and for many pupils dry) traditional science topics.

A potential philosopher's stone

Arguably, what recent experiments in prescribing science curriculum have shown is the importance of achieving a balance between teaching about (1) the products of science, and (2) the processes by which initially creative ideas are tested (and the evidence debated) until they are accepted into the pantheon of scientific knowledge – that body of knowledge we consider robust and yet always provisional. In particular, there is immense scholarship now on how the history and philosophy of science can inform teaching and learning about science in ways that help learners understand the nature of science itself (Hodson 2009; Matthews 1994).

Teaching about the nature of science is not an elixir that will solve all our educational problems and lead to an immortal and immutable science curriculum. However, arguably, the key to public scientific literacy is appreciating the epistemology of science: Why is it these scientists say they cannot get absolute proof, yet they want to tell us what we should do? And why do they sometimes still disagree after they have done the 'experiments' and seen the evidence? Moreover, many learning difficulties in science subjects relate to a failure to appreciate the ontology of science – what a theory, a model, a law etc. actually are – and so why we teach about electron orbits if we 'know' they do not exist (Taber 2010). Work with gifted science students suggests that a focus on the nature of science can be an effective way to make science more challenging for the most able (Taber 2007a). Of course, differentiation is still a major issue: the full ability range will not benefit from the same level of treatment of historical case studies or philosophical arguments about scientific method any more than all will learn optimally from the same treatment of current electricity or genetics. But teaching about the nature of science has much to offer in preparing both a scientifically-literate citizenship and the high-flying future research scientists the world needs.

Conclusion

Ultimately 'science for all' cannot mean treating all students the same: it needs to recognise levels of motivation and attainment, and differences in the particular interests, skills and aptitudes of different students. Many students are gifted in some aspects of science learning, just as many – even the highest attainers – have learning difficulties in some aspects of the curriculum. A major area of research in science education makes it quite clear that no two students will actually experience the same teaching and so undergo the same learning, even when sitting next to each other in the same lesson (Taber 2011), so pretending that *being fair* means everyone gets presented with the *same fare* is actually disingenuous.

The challenge is to develop a curriculum framework that can *in principle* offer the same opportunities for all, while being differentiated to meet the needs of the most gifted – but also everyone else. Within that framework, some will meet highly-abstract, theoretical and mathematicalised teaching; others will focus more on exploring the science involved in important contexts, perhaps from a more grounded and personally-relevant perspective. However, authentic science education for any student cannot be devoid either of concepts or issues, neither of theory, nor authentic investigative (and sometimes extended) practical work. The real question is: *To what extent* does the teaching of these different elements need to be differentiated and customised for different groups of learners?

Key questions

1   Is science education too important to be tightly prescribed by amateurs such as government ministers who often have limited qualifications in science and no experience of classroom teaching? Would science education instead fare better if teaching were treated as a genuine profession, where professionals use their expertise to make decisions on such matters as curriculum content?

2   To what extent are the problems discussed in this chapter likely to have universal solutions, rather than needing different approaches in (for example) the education systems of wealthy, developed and poor, developing countries?

3   To what extent does the school subject 'science' both tend to down-play the very real differences across science disciplines and limit the opportunities (1) for students to develop specialist interests and exercise choice in their study of the sciences, and (2) for teachers to teach to their strengths?

Further reading

If the issues raised in this chapter have given you cause to think about your own ideas about ability in science and the aims and means of teaching science 'for all', then you may find the following readings of interest.

Is there still a need for gifted education?

In a paper from the journal *Learning and Individual Differences*, Reis and Renzulli (2010) review the literature on provision for gifted learners and argue that the 'need for programs and services for [our most able students] may be more critical than in any time period in recent history' (ibid.: 316).

Why are modern scientists so dull?

In an editorial in the journal *Medical Hypotheses*, Charlton (2009) argues that the modern education system tends to select scientists as much for their conscientiousness and agreeableness as for high intelligence and creativity. Yet, Charlton argues, scientific progress depends particularly on the latter characteristics. He concludes that 'modern top scientists are likely to be less intelligent and creative than is desirable, and probably significantly less intelligent and creative than top scientists used to be' (ibid.: 242).

Is school science a questionable construct?

In a paper in the *Journal of Curriculum Studies*, Jenkins (2007) outlines some of the history of the notion of school science. Jenkins argues (reflecting a strong theme in the philosophy of science) that there is no simple 'scientific method' common across science disciplines, and that science education often fails to acknowledge 'the major conceptual and philosophical differences that exist between the sciences' (ibid.: 273).

References

Assocation for Science Education (1981) *Education through Science: Policy statement.* Hatfield, Hertfordshire: The Association for Science Education.

Cerini, B., Murray, I., and Reiss, M. (2003) *Student review of the science curriculum: Major findings.* London: Planet Science/Institute of Education/Science Museum (p. 33).

Charlton, B. G. (2009) Why are modern scientists so dull? How science selects for perseverance and sociability at the expense of intelligence and creativity. *Medical Hypotheses,* 72(3): 237–43. doi: 10.1016/j.mehy.2008.11.020

Department for Education (2011) Michael Gove to Cambridge University (25 November). Available online at: http://www.education.gov.uk/inthenews/speeches/a00200373/michael-gove-to-cambridge-university (accessed 25 June 2013).

Department of Education and Science & Welsh Office (1985) *Science 5–15: A statement of policy.* London: Her Majesty's Stationery Office.

DfEE/QCA (1999) *Science: The National Curriculum for England, Key Stages 1–4.* London: Department for Education and Employment/Qualifications and Curriculum Authority.

Emsley, J. (2010) *A healthy, wealthy, sustainable world.* Cambridge: RSC Publishing.

Gilland, T. (ed.) (2006) *What is science education for?* London: Academy of Ideas.

Gould, S. J. (1992) *The mismeasure of man.* London: Penguin.

Hodson, D. (2009) *Teaching and learning about science: Language, theories, methods, history, traditions and values.* Rotterdam, The Netherlands: Sense Publishers.

Hudson, L. (1967) *Contrary imaginations: Psychological study of the English schoolboy.* Harmondsworth: Penguin.

Jenkins, E. W. (2007) School science: A questionable construct? *Journal of Curriculum Studies,* 39(3): 265–82.

Matthews, M. R. (1994) *Science teaching: The role of history and philosophy of science.* London: Routledge.

Millar, R., and Osborne, J. (1998) *Beyond 2000: Science education for the future.* London: King's College.

Osborne, J., and Collins, S. (2001) Pupils' views of the role and value of the science curriculum: A focus-group study. *International Journal of Science Education*, 23(5): 441–67. doi: 10.1080/09500690010006518

QCA (2007a) *Science: Programme of study for Key Stage 3 and attainment targets.* London: Qualifications and Curriculum Authority.

QCA (2007b) *Science: Programme of study for Key Stage 4.* London: Qualifications and Curriculum Authority.

Reis, S. M., and Renzulli, J. S. (2010) Is there still a need for gifted education? An examination of current research. *Learning and Individual Differences*, 20(4): 308–17. doi: 10.1016/j.lindif.2009.10.012

Snow, C. P. (1959/1998) *The Rede Lecture, 1959: The two cultures.* Cambridge: Cambridge University Press (pp. 1–51).

Statutory Instrument (1989) The Education (National Curriculum) (Attainment Targets and Programmes of Study in Science) Order 1989. London: Her Majesty's Stationery Office.

Taber, K. S. (2007a) *Enriching school science for the gifted learner.* London: Gatsby Science Enhancement Programme.

Taber, K. S. (ed.) (2007b) *Science education for gifted learners.* London: Routledge.

Taber, K. S. (2008) Towards a curricular model of the nature of science. *Science and Education*, 17(2–3): 179–218. doi: 10.1007/s11191-006-9056-4

Taber, K. S. (2009a) Learning from experience and teaching by example: Reflecting upon personal learning experience to inform teaching practice. *Journal of Cambridge Studies*, 4(1): 82–91.

Taber, K. S. (2009b) Maths should be the last thing we teach. *Physics Education*, 44(4): 336–8.

Taber, K. S. (2010) Straw men and false dichotomies: Overcoming philosophical confusion in chemical education. *Journal of Chemical Education*, 87(5): 552–8. doi: 10.1021/ed8001623

Taber, K. S. (2011) Constructivism as educational theory: Contingency in learning, and optimally guided instruction. In J. Hassaskhah (ed.) *Educational Theory.* New York: Nova (pp. 39–61). Available online at https://camtools.cam.ac.uk/wiki/eclipse/Constructivism.html (accessed 25 June 2013).

Taber, K. S. (2012) Meeting the needs of gifted science learners in the context of England's system of comprehensive secondary education: The ASCEND project. *Journal of Science Education in Japan*, 36(2): 101–12.

Taber, K. S. (2013) *Classroom-based research and evidence-based practice: An introduction* (2nd edn). London: Sage.

White, J. (1987) The comprehensive ideal and the rejection of theory. *British Journal of Educational Studies*, 35(3): 196–210.

Chapter 3

Government policy

Can short-term strategies bring long-term change?

Chris Lloyd-Staples

Introduction

I can well remember my first day as a teacher in 1979. My Head of Department gave me my 'scheme of work' for the year. 'Well,' he said, handing me a scrap of paper, 'in the first year we do the human body, and in the second year we do types of plants and animals. The third year is processes, and then we start the O level and CSE. Decide for yourself what you teach in the lower school.' Baptism of fire. And this was a very respected school in the area. Was this typical, I wondered? Without a central government policy or framework, each school functioned in isolation.

For the next decade, nothing profoundly changed in science teaching, or so it seemed to me with my daily existence located in classroom practice. As a member of the Association for Science Education (ASE) and the Institute of Biology, I was well aware of issues in science teaching, but my school appeared autonomous and driven by the needs of the children. Certainly, we were unaffected by academic researchers or by the Government of the day.

In 1988, things changed suddenly with the National Curriculum and the GCSE. Suddenly there was a sense of greater minds thinking of what would be best in education, and steering the system by imposing an overall direction. Despite grumbles about changing our established practice, there was an underlying acceptance of the need to rationalise what was being taught.

I left the classroom in 2000 to become a county adviser for science. After two decades in the teaching profession, I could see that teaching processes had become somewhat more child-centred, and scientific content had become more relevant and sequential. However, there had been no radical transformation, and general classroom activity remained much as it had always been. Importantly, pupil attitudes to science were recognised as being negative overall, with the subject seen as difficult and un-engaging (Coe *et al.* 2008).

The introduction of the National Strategy

The National Strategy for Key Stage 3 Science was one of a number of National Strategy initiatives that were introduced into schools from 1998, beginning with Literacy and Numeracy at primary level, with a phased expansion to cover Core and Foundation subjects up to Key Stage 3, and eventually to include Key Stage 4. These new initiatives included new approaches from the Department for Education and Skills/Department for Children, Schools and Families (DfES/DCSF), and new syllabus and assessment models from examination boards, under direction from the Qualifications and Curriculum Authority (QCA).

However, within a few years of the initial introduction of the National Strategies, countless other externally-imposed 'whole-school' initiatives impacted on the way that schools operated. All of these distracted schools from concentrating on the vital messages of the Strategies. It would be difficult to mention all of the initiatives and changes, but significant examples included, amongst others:

- Assessment for Learning
- Assessing Pupil Progress
- Every Child Matters
- Condensed Key Stage 3
- Personal Learning and Thinking Skills
- Big Picture of education
- Changes to GCSE and A-level syllabuses
- 14–19 Diplomas
- Outdoor Classroom
- One-to-one tuition
- Expanded use of ICT (Information Communication Technology)
- Workforce reform
- Dropping Key Stage 3 National Tests (SATs)
- National Challenge and Gaining Ground schools
- School Improvement Partners
- Specialist School status
- Self-Evaluation Form.

In addition to these general changes, in science teaching there was a focus on aspects of:

- Level 6 Plus
- Triple Science
- Core and Additional Science at GCSE
- Applied Science
- Science Learning Centres
- How Science Works.

Each of these initiatives, rolled out with promotional material and funding, involved schools in considerable effort to implement the required changes. While some of the changes were described as 'non-statutory', the majority of schools adopted each in turn as the school was encouraged to conform and make use of available funding. The various initiatives were disruptive to the school organisation to a greater or lesser extent, and invariably saddled classroom teachers with additional work in order to adapt their practice to meet new demands. Typical teaching plans became ever more complicated, with boxes to record how the lesson would cover each component of content or element of assessment.

The pressure to adopt the initiatives came from central government, but was also amplified via local authorities and through competition with neighbouring schools. However, many of the new initiatives were hastily introduced after a brief but well-funded pilot trial in chosen schools (e.g. Hallam *et al.* 2006; DCSF 2009; DCSF 2010). The evidence for the long-term benefit of each initiative was often sketchy at best, and many turned out to be expensive failures, as in the case of the 14–19 Diplomas. After huge investment of time, effort and funding, each initiative was dropped following loss of funding and political support.

The introduction of new initiatives was not (publicly at least) associated with clear explanation of the outcomes expected from the initiative. While schools had overall targets, the impact of the initiative at a local level, and separated from the impact of other concurrent initiatives, was never clear to the teachers involved in the classroom. As Friedland terms it, the initiatives had no Results Accountability at a local level:

> If we knew what outcomes we wanted and how to measure them, the next logical thing to do would be to assess how we are doing on each of the measures. We would then forecast … whether things were likely to get better or worse if we just kept doing what we were doing.
>
> (Friedland 2005: 6)

Although any one individual teacher may not be involved in all of these initiatives, the cumulative impact of the changes – creating a constant sense of turmoil and uncertainty – caused a sapping of energy and less enthusiasm for innovation (Ofsted 2010; see also BBC News 2010).

> … the sheer number of initiatives and programmes and the speed at which schools are expected to implement them may be counterproductive
>
> (OECD 2008: 144)

In addition, the impact of this overload often resulted in two outcomes, whereby schools effectively act to minimise the trauma of seemingly endless change:

- Superficial and temporary engagement with the initiative, such that the changes are not embedded or developed to their full potential. The initiatives are often dropped or at least reduced as soon as another initiative is introduced.
- Box-ticking: providing minimal evidence that the initiative has been 'done' in order to satisfy monitoring bodies and obtain necessary funding.

The National Strategy continued to be the greatest influence, and became the delivery agent for many of the subsequent initiatives. The Strategy initially concentrated on subject strands in Key Stage 3, but in 2003 the focus shifted to whole-school approaches (see DfES 2003a, 2003b), and later the National Strategy expanded to include Key Stage 4.

The intended purpose of the National Strategy for Science

The National Strategy for Key Stage 3 was introduced in 2002 in response to recognised problems in Key Stage 3 (Years 7–9, ages 11–14). These problems were perceived to be a lack of progression and loss of motivation in many students of this age, with particular worsening towards the end of National Curriculum Year 7 and throughout Year 8 (see Omerod and Duckworth 1975; Ruddock *et al.* 2004). The disappointing performance at Key Stage 3 had been highlighted by improvements in Key Stage 2:

> The success at primary level over the last two years has brought into sharp focus the unacceptable lack of progress from age 11 to 14.
>
> (Blunkett 2000)

Ryan (2002) discussed some of the current understanding of the reasons behind this dip. She identified factors that may play a part in the overall problem at transition:

- changes in curriculum structure – new subjects and different teachers;
- changes in the style of language used;
- changes in teaching and learning styles;
- changes in environment – both the school itself and the learning base;
- time constraints in secondary schools;
- variation in primary schools, requiring topics to be retaught to secondary classes;
- lack of awareness by secondary teachers of what is taught at primary school;
- changes in attitudes and motivation during adolescence.

The DfEE (later DCSF) had long recognised the existence of a problem within Key Stage 3 (the 'forgotten Key Stage'). The causes of these issues were numerous and interwoven, but the central issues were felt to be a lack of planned progression in schemes of work, and overly-didactic teaching methods. As a result, pupils who had made good progress through their primary schools were suddenly encountering traditional teaching methods that lacked individual engagement, and were often repeating work with which they were already familiar. Thus, '… the dominant experience in Year 7 is really one of repetition of work rather than true progression, as some research suggests …' (Braund and Driver 2002: 5). Faced with this, it was widely recognised that students were becoming disillusioned and sometimes poorly behaved.

Research commissioned by the DfEE helps explain this poor performance. Achievement slips in the transition from primary to secondary schools. Too little is expected of pupils in the first year of secondary school, by the end of which around a third of pupils perform worse in tests than they did a year earlier.

(Blunkett 2000)

What did the Science Strategy look like for teachers?

The initial phase of introduction included the publication of a Framework for Key Stage 3, which provided a structure for teaching built upon a fundamental statement of good practice, and new thinking about the development of key scientific concepts through a progression of stages. The document was introduced to schools through carefully-staged training events, to which a representative from each school was invited. The intention was that the information would be cascaded back to the science teachers, and the Framework implemented.

Further support for schools was delivered by consultants, who provided a sequence of training sessions, as well as in-school support working with individual teachers and departments. Indeed, the volume and frequency of training began to be a problem in itself. Engagement with the strategy had been close to 100 per cent at the outset, but this gradually declined over the years. Each training event was accompanied by extensive resources in white Strategy folders, and after each training session it was intended that these folders would form the basis of cascaded briefings in each school, delivered by the participant who had attended the central training. Needless to say, it was commonplace to see rows of these white folders on the shelves in school science departments, underused and gathering dust.

The training sessions themselves were very well received and the delivery was generally of excellent standard. The training materials were equally well designed, and the content was seen as being both useful and relevant to the changes in science teaching. The content contained ideas for the improvement of pedagogy, as well as subject-specific sessions on how to put across difficult ideas. The provision of these new approaches would reach a high point around 2004, and gradually the emphasis shifted towards encouraging science teachers to evaluate their work. Instead of reporting on their delivery of training, the strategy consultants were increasingly asked to report to the government on the impact of their work.

Outcomes: the national costs and benefits of the National Strategy

The expenditure on the National Strategies was immense. The overall annual figure, taking into account the gradual shift in the focus of the Strategies and the expansion to include other subjects and age-groups, is recorded in Table 3.1.

Table 3.1 Annual expenditure on the National Strategy (£m.)

Year	Expenditure
2002–03	425
2003–04	465
2004–05	335
2005–06	410
2006–07	460
2007–08	470
2008–09	420
2009–10	450

Source: DCSF.

The issue reported in many schools was that, when the focus shifted from one initiative to the next, the previous initiative declined in importance and many of the advances were lost. The lack of sustained pressure on a small number of proven 'high-impact' strategies appears to have led to a general failure to make progress in England and Wales. This lack of progress can be seen in the limited change in the examination outcomes since the introduction of the National Strategy for Science (see Table 3.2).

Thus, in the years since the initial launch of the National Strategy for Science in Autumn 2002, the performance of students increased 3 per cent at Level 5+ and 1 per cent at Level 6+ between 2003 and 2008. Given that the Strategy would have taken time to have an impact on teaching and hence results, the change attributable to the Strategy effect is less than this, possibly amounting to zero.

It is possible that the contribution of the National Strategy may be seen in areas other than examination results, for example in promoting student interest

Table 3.2 Percentage of students achieving Level 5 and 6 in National Tests (SATs) at Key Stage 3 (age 14)

Year	Science		English		Maths	
	Level 5+	Level 6+	Level 5+	Level 6+	Level 5+	Level 6+
2002	67	34	67	32	67	45
2003	68	40	69	34	71	50
2004	66	35	71	34	73	52
2005	70	37	74	35	74	53
2006	72	41	73	34	77	57
2007	73	40	74	33	76	56
2008	71	41	73	33	77	57

Source: National data releases.

in science, encouraging uptake post-16, and supporting the work of less-qualified teachers in a period of staff shortages. The impact (or otherwise) of the Strategy cannot be judged by examination results alone.

Evaluations have tended to be in the form of annual reports, often focused on the apparent stagnation of national examination results in all three Core Subjects. The National Literacy Strategy (NLS) and the National Numeracy Strategy (NNS) were evaluated in 2002 (DfES 2003c), at the time that the Science Strategy was being introduced. This report noted that results in Literacy and Numeracy had improved, but '... much of the increase occurred prior to the introduction of the NLS in 1998 and NNS in 1999 ...' (ibid.: 3). The same report noted that:

> Although the Strategies have made a good beginning in a relatively short period of time, the intended changes in teaching and learning have not yet been fully realised. After four years, many see NLS and NNS as needing to be re-energised; the early momentum and excitement have lessened and a new boost would be helpful.
>
> (Ibid.: 8–9)

This being the case after four years, it was important to ask whether the same issues were true in science – whether the early impact in 2002–2004 in changing teaching styles was indeed sustained over the remaining duration of the Science Strategy without a re-energised approach, and where other initiatives gradually took taken centre-stage.

The wider question, therefore, was whether embedded changes in education can be driven by short-term initiatives, or whether systemic change requires a sustained focus on an objective for possibly a decade, in order to overcome the inherent resistance of the status quo.

The end of the National Strategies

The National Strategies ended in 2011, which for the Science component was nine years after the introduction and initial roll-out. The decision to end the programme was motivated by two factors: the economic downturn made the financing of the Strategy hard to afford, and the stagnation of the results has made the Strategy hard to justify (see Table 3.2). This decision to terminate the funding for the Strategy had been anticipated, as the Strategy was always seen as time-limited. Schools had increasingly come to depend on external support from Strategy Consultants, and it was the loss of this external support, rather than the loss of the Strategy itself, that schools came to regret.

Against this backdrop, a number of decisions had already been taken which would suggest to an observer that science, in particular, had become less important in the school curriculum. For example, science lost its external national testing at Key Stage 2 (Year 6, age 11) from 2010 and testing had already been dropped in Key Stage 3 (Year 9, age 14) in 2009. All achievement targets that included science were dropped, and the remaining target was the number of students gaining five or more GCSE passes at A*–C grades, including English and Mathematics. At all levels, the importance of English and Mathematics was stressed, and Science barely registered as a 'Core Subject' in the curriculum.

Despite this, all political parties have indicated their continued commitment to science, and the need to produce new generations of qualified scientists. The numbers of students taking A-level courses in mathematics, chemistry and physics were monitored as part of the reporting process between local authorities and the Government (see DCLG 2007).

It had become quite clear that the National Strategies failed to lead to a breakthrough in examination outcomes and, as such, the commitment of significant funding and effort could be seen as doing nothing more than arresting further decline in attitudes towards science. However, this begs the question about the introduction of initiatives without sustained drive and without considering the impact at a local level – in this case at school level.

What were the school-level outcomes?

With such huge investment in the National Strategy for Science, it is reasonable to ask whether the intended transformation did indeed take place. After so much funding and human investment at national, local authority, school and individual teacher level, it would be expected that science lessons demonstrated significant improvement, with raised student outcomes. At the very least, surely, students would be more engaged with science and more positive in their study of the subject.

The evidence suggests that this is simply not the case (Lloyd-Staples 2012). A science lesson today would in general look and sound very much the same as a lesson three decades ago. Other than an increased use of ICT to support the delivery, as well as other general changes in pedagogy, there are few changes that can be recognised and directly attributed to the National Strategy (Ofsted 2010). In some areas the general National Strategy can be seen to have made a difference, and there is now much more discussion of processes within teaching, alongside sharing of ideas and resources. Whereas in the past, when teachers worked as individuals, preparing their own material and teaching behind closed doors, science departments are now more open to teamwork. Teaching methods are open to scrutiny, and good practice is spread within the science team and beyond. One of the key ideas of the Strategy – a three-part lesson – has become well-established in most schools. The use of lesson objectives and the use of formative assessment are now routine.

As for the elements of the Secondary Science Strategy that are specifically subject-related, there is little to be seen in current practice. The white Strategy folders, full of new approaches and ideas, have long vanished from most schools. On this level, it is as if the Strategy had never happened.

Why did the Strategy fail to embed?

With hindsight it may be easy to recognise the features of the National Strategy which led to its failure to have a profound and lasting legacy. First, as with all top-down approaches, there was a need to engage the involvement of the whole profession with an agreed sense of direction and urgency to change. This process began well, but increasingly schools began to see the Strategy as imposing a set of approaches which may or may not work in particular schools. Over the duration of the Strategy, schools became progressively less interested in the Strategy message and more interested in using consultant support to develop teaching to suit the needs of the school. In particular, high performing schools would often see the National Strategy as an irrelevance.

The second problem was the dependence on the cascade method for spreading messages back into schools (Hayes 2000). Despite excellent central delivery of Continuous Professional Development (CPD) by Strategy consultants, the problems facing participants when they returned to their own school were immense. For even the most motivated teacher, there were huge difficulties in condensing a full day of training into a package that would engage their colleagues and persuade them to adopt new ideas. With huge day-to-day workloads, most schools found it difficult to provide time for a teacher to train their colleagues. After a few weeks, the ideas became gradually diluted and lost, and the impact of the training was minimal. Schools began to recognise that releasing teachers for training was proving expensive and most damaging to teaching continuity. As a result, attendance at training events began to decline.

The most serious barrier faced by the National strategy was the increasing sense of initiative overload. This phenomenon has been recognised as a problem facing British schools during this period, and resulted from new government policies and initiatives arriving in schools with a frequency that proved ultimately counter-productive (OECD 2008). While each new idea was in itself a valuable contribution, the latest initiatives were inevitably a distraction which drew teachers away from what had gone before. There was increasing despair, and schools began to pay lip service to the newest initiatives while allowing previous priorities to fade.

What lessons can we learn?

It appears that the Strategy failed in its aims because it failed to listen to its own message. It failed to recognise that teachers, just as much as students, need simple messages, repeatedly delivered in innovative ways, in order to learn and fully internalise these ideas. These issues were summarised by a respected Head of Department in 2004:

> I think that the whole Strategy was over-ambitious. I mean, if you think about eight talks, you know, with twelve, fifteen bullet points in each, a hundred points ... it was way too ambitious in the amount of material they

expected people to cover. It would have been better if they'd said 'These are the three points to take away from this. Please now, try and implement these three.' It didn't seem to have been thought through, and it seemed to be almost a shotgun approach rather than a bullet approach ....

Thus, it appears that the sheer volume of messages became too much for schools to absorb. Teachers returning from training with a dozen new ideas to promote within the school would soon find that they and their colleagues were inundated. There were so many ideas that teachers simply could not implement what was being suggested, let alone embed it into their ongoing practice.

Depending on the Cascade method (see Morrison *et al.* 1989) of feeding back from Strategy training was also a difficulty, as some teachers did this well, advocating changes in department practice, while others were much less effective. This situation was summarised by the same Head of Science in 2004:

> ... some people have given detailed feedback, others just one or two points, like the one who went on 'Investigations', we had the post-it type sheets, which were quite handy, but not a lot else. Whereas the one on 'Assessment' that I attended, there was a bullet by bullet, point by point report covering some of the things like 'no hands-up', and no grades on work, and so on. There was quite a difference.

It can be seen here that, in addition to the differences in detail of the feedback to the school, it is clear that the ideas were being selectively implemented. In other words the schools were making value judgements about which parts of the strategy they would choose to implement. It is also evident that feedback from a senior member of the team is more likely to lead to significant change in practice than would be the case if the central training event had been attended by a more junior teacher in the school.

The centrally-imposed National Strategy also under-estimated the extent to which schools, and departments within them, operate as communities of practice (Wenger 1998). Within these teams, effectively isolated from teachers in other schools on a day-to-day level, there had developed a firmly-established ethos and a set of procedures. In addition to the resistance to change that is inherent in any system, science teachers were slow to embed these new ideas because, individually, they were unaware of the wider momentous changes taking place within science teaching.

These difficulties could have been addressed by relentless concentration on a few key messages and the repeated emphasis on this group of central ideas. But this was not to be. Instead, each of the numerous central training events added yet more ideas for schools to adopt, and as a result science teachers adopted some and ignored the majority. In consequence, despite a veneer of change, very few of the pedagogic or subject-related ideas from the National Strategy have been fully embedded into everyday practice in schools.

Student attitudes to science

The ultimate judge of any change in teaching would be seen in improved student outcomes, or at the very least in improved attitudes towards the subject, leading to career or education choices. It is evident that during all stages of the National Strategy there was no significant change in examination outcomes or selection of science A-levels. Is there any evidence to suggest that students came to hold more positive views about science as a subject and as a career pathway?

A small-scale survey conducted in 2008, involving students taught during the heyday of the Science Strategy, suggested that little or no change had taken place (Lloyd-Staples 2012). Their perceptions of scientists remained stubbornly negative and stereotypical:

> White coat, big glasses, funny hair ... 'cos when you're a child and you watch cartoons, and things, there's a scientist they're always like this crazy science person.
>
> (Year 7)

> ... all scientists are different. Like – a scientist could be like, inventing biological weapons and a scientist could be like inventing a cure for like cancer. How do I say it ... some of them are good, like, and some of them are bad.
>
> (Year 9)

> A man, in a white jacket, with glasses on, looks like a right nerd ... That's Einstein, frizzy hair. You think of someone who actually has no social life, that's what I do think of.
>
> (Year 11)

In addition, the subject was often seen as difficult, not engaging, and irrelevant to daily life. After great anticipation about studying science in secondary school, a Year 7 student commented:

> ... the last experiment we did was actually quite boring 'cos all we had to do was look at rocks. Dip them in water and see what colour they changed ....

One student after another expressed disappointment about the nature of science that they were being taught, and the way that the subject was being put across:

> I think science in general is actually quite a boring subject. I think that it could be made quite fun. I don't think teachers here make it fun, at all. I just think it's 'read out of a book and learn it that way', and that's about as far as it goes.
>
> (Year 11 student)

In other words, at least for the students sampled in this survey, the National Strategy had failed to impact upon their views of the subject or their feelings about scientific careers, and it appears that the teaching of the subject remained unengaging. Indeed, the evidence of surveys in 2004 and 2008 indicated that the use of National Strategy techniques had peaked in the early intense period of the Strategy, and that, instead of further embedding, the methods went into rapid and selective decline, with some Strategy ideas no longer used at all.

Conclusion: reflections on moving from government policy to established classroom practice

The National Strategy for Science illustrated many of the difficulties in moving from a nationally-imposed framework to an established approach, despite the concepts being soundly composed, fully-funded and well-received by the profession. In the event, for these ideas to become embedded into everyday routine practice, the pace of introduction needed to be slower, with fewer new key concepts to incorporate at any time. Fundamentally, these key ideas needed to be repeatedly presented, reinforced and set into contexts that all science teachers could recognise as fundamental to their own teaching. Above all else, it was essential for there to be no further distractions to draw teachers away from the process of embedding the new approaches into their practice.

In effect, the Government had been too ambitious and had attempted to change too much, too quickly. With clearer consideration of the key aims and the process of delivery, these aspirations could have been achieved.

As it was, in hoping to change student attitudes, the Strategy also ignored the powerful influences of wider society, including the impact of the media, perceived employment prospects and peer pressure. All of these were revealed to have far greater influence on teenagers than improved teaching could ever hope to achieve. Without a fundamental societal change in the way that the work of scientists is valued and appreciated, students have remained generally neutral, at best, with regards to science. Let us hope that parents and friends begin to encourage children in their choice of science as an option, rather than trotting out their usual comment that science is hard, boring, and they hated it at school.

Is it possible to move from policy to practice, changing student attitudes and improving science results? The answer is surely 'Yes!', but the policy-makers need to consider more carefully how to achieve this transformation.

Key questions

1   The National Curriculum was first introduced in England, Wales and Northern Ireland in 1988 to counter the variability in what was being taught. The new proposed National Curriculum for 2014 onwards can be disapplied by Academy schools, and Local Authority schools have been given permission to be creative in their interpretation of the curriculum. What are the implications of this new variability between schools, and how may science teachers be affected by this new policy?

2   Current thinking from the Department for Education in Britain proposes that schools are effective innovators and that local school-to-school support is the most effective way to spread good practice from one institution to another. What are the implications of developments taking place at the level of small clusters of schools, rather than at a regional or national level?
3   New policies within the National Curriculum and within examination specifications are leading to: (1) a greater emphasis on factual knowledge rather than on transferable skills; and (2) linear courses with all assessment in the form of a terminal examination. To what extent may these changes impact on student enjoyment of science and their attitudes towards the subject?

Further reading

BBC News (2010) *Ofsted criticises Three Rs 'initiative overload'* (24 February). Available online at: http://news.bbc.co.uk/1/hi/education/8530575.stm (accessed May 2010).

National Strategy archives. Available online at: http://www.education.gov.uk/schools/toolsandinitiatives/nationalstrategies.

National Strategy for Science. *Framework*. Available online at: http://www.nationalstemcentre.org.uk/dl/cc25f54c9695a693c2d811cc491c-5c53d28ac6df/17087-2002%20Science_Framework.PDF.

The Importance of Teaching: The Schools' White Paper 2010. Available online at: https://www.education.gov.uk/publications/standard/publicationdetail/page1/CM%207980. This document describes current and future government policy, and outlines the shift from centralised approaches to local self-improving systems. There are many proposals which directly or indirectly impact on the status of science teaching, not least being: 'We will continue to provide additional support for the uptake of mathematics and the sciences. A strong national base of technological and scientific skills is essential to growth and employers continue to report shortages of these skills.'

Waters, M. (2013) *Thinking Allowed on Schooling*. Carmarthen: Independent Thinking Press. This book gives a fresh view of the relationship between external demands – such as government policy – and the operation of schools. This view is controversial because it contrasts official policy and the pragmatic responses which take place in schools. In this respect, the book discusses many of the issues raised by the implementation of the National Strategy.

References

BBC News (2010) Ofsted criticises Three Rs 'initiative overload' (24 February). Available online at: http://news.bbc.co.uk/1/hi/education/8530575.stm (accessed 25 June 2013).

Blunkett, D. (2000) *Raising Aspirations for the 21st Century*. Speech to the North of England Education Conference, Wigan, 6 January. London: DfEE.

Braund, M., and Driver, M. (2002) *Moving to the Big School: What do Pupils Think About Science Practical Work Pre- and Post-transfer?* Paper presented at the British

Educational Research Association (BERA) Annual Conference, University of Exeter, 11–14 September.

Coe, R., Searle, J., Barmby, P., Jones, K., and Higgins, S. (2008) *Relative Difficulty of Examinations in Different Subjects, Summary Report 2008*. Durham: Durham University, CEM Centre.

DCLG (2007) *The New Performance Framework for Local Authorities and Local Authority Partnerships: Single Set of National Indicators*. London: Department for Communities and Local Government.

DCSF (2009) *The National Challenge – Raising Standards, Supporting Schools: Gifted and Talented Pilot Programme* (Introduction and Overview). Ref: 00390-2009. Nottingham: Crown Copyright.

DCSF (2010) *Evaluation of the Making Good Progress Pilot*. Research Report RR184. London: DCSF/Pricewaterhouse Coopers.

DfES (2003a) *Key Messages from the Key Stage 3 National Strategy*. London: Department for Education and Skills.

DfES, (2003b) *Preparing for Change: Evaluation of the Implementation of the Key Stage 3 Strategy Pilot*. 0208/2003. London: Department for Education and Skills.

DfES (2003c) *Watching and Learning 3: Final Report of the External Evaluation of England's National Literacy and Numeracy Strategies*. Ontario Institute for Studies in Education [OISE/UT]. 0101/2003. Nottingham: Department for Education and Skills.

Friedland, M. (2005) *Trying Hard is Not Good Enough: How to Produce Measurable Improvements for Customers and Communities*. Victoria, BC: Trafford.

Hallam, S., Rhamie, J., and Shaw, J. (2006) *Evaluation of the Primary Behaviour and Attendance Pilot*. London: Department for Education and Skills.

Hayes, D. (2000) Cascade Training and Teachers' Professional Development. *ELT Journal*, 54(2): 135–45.

Lloyd-Staples, C. (2012) *Implementation and Impact of the Secondary Science National Strategy*. Saarbrücken: Lambert Academic Publishing.

Morrison, K., Gott, R., and Ashman, T. (1989) A Cascade Model of Curriculum Innovation. *Professional Development in Education*, 15(3): 159–69.

OECD (2008) *Improving School Leadership, Vol. 2: Case Studies on System Leadership*. Paris: OECD. Available online at: www.sourceoecd.org/education/9789264044678 (25 June 2013).

Ofsted (2010) *The National Strategies: A Review of Impact*. Ref. 080270. London: Crown Copyright.

Ormerod, M. B., and Duckworth, D. (1975) *Pupils' Attitudes to Science*. Slough: NFER.

Ruddock, G., Sturman, L., Schagen, I., Styles, B., Gnaldi, M., and Vappula, H. (2004) *Where England stands in the Trends in International Mathematics and Science Study (TIMSS) 2003: National Report for England*. Slough: NFER. Available online at: www.nfer.ac.uk/timss2003 (accessed 17 February 2007).

Ryan, M. (2002) Tackling Key Stage 2 to 3 Transition Problems: A Bridging Project. *School Science Review*, 84[306]: 69–76.

Wenger, E. (1998) *Communities of Practice: Learning, Meaning, and Identity*. Cambridge: Cambridge University Press.

Chapter 4

Girls in science: who puts up the barriers?

Joanne Cole

Introduction

There is a strong and increasing demand for people skilled in science, technology, engineering and mathematics (STEM) across the UK economy. Despite this need, there are still only a limited number of people with these skills entering the job market. An education and skills survey carried out by the Confederation of British Industry (CBI 2011) shows that almost half of employers questioned were already struggling to source suitably-skilled employees, with that fraction expected to increase during the following three years.

It is an equally well-known problem that, in Western Europe, women are underrepresented in STEM subjects, particularly in areas such as physics, with the United Kingdom being one of the worst offenders; women represent only 12.3 per cent of people working in SET occupations, compared to 45.1 per cent of the overall UK workforce (Kirkup *et al.* 2010).

The under-representation of women in science is further highlighted by recent A-levels results: Figure 4.1 shows the percentage of girls and boys who sat an A-level in physics, chemistry or biology in the period 2007–2012. It is clear that, while there are systematically more girls than boys studying biology at A-level and systematically fewer studying chemistry, the numbers are not as dramatically different as they are for physics, where there are more than four times as many boys studying physics at A-level as there are girls.

This issue has been acknowledged as an important one at all levels, with the UK Prime Minister David Cameron commenting at a Q&A session in 2010 that 'if we think that somehow all the good scientists are men, we are making a big, big mistake' (UKRC 2010). The clear implication of these facts is that the UK economy is suffering because we are failing to engage more young people – in particular more young women – with STEM subjects such as physics.

So, clearly we have a big problem. Right?

But what is the problem?

But what if the reason for the under-representation of women in STEM is simply that girls are not as good at science as boys? That's possible, surely? Indeed, in

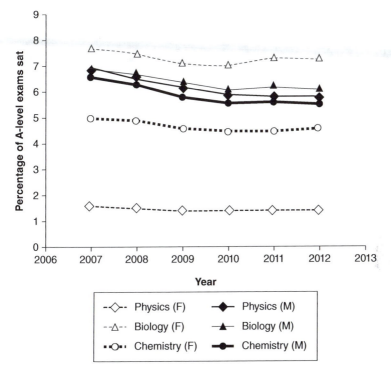

Figure 4.1 Percentage of A-levels sat in physics, chemistry and biology for the period 2007–2012.

Source: Based on figures from the Joint Council for Qualifications.

2005 the then President of Harvard University, Dr Lawrence Summers, suggested exactly that: that maybe the reason why there are relatively few women in high-level STEM careers is down to differences in the innate mathematical ability between girls and boys.

Unsurprisingly, Dr Summers' suggestion, aired at a private conference in the United States, provoked a furore among many female academics, despite the fact that he subsequently explained that he was taking a deliberately provocative position in his presentation. In response to Dr Summers' remarks, a paper was published which looked into the extensive body of literature available addressing this issue (Halpern *et al.* 2007); it came to the conclusion that there is no simple answer to the question of whether the difference in innate mathematical ability between the genders is the reason for the lack of women in STEM subjects.

Studies reported in the paper show that boys do indeed tend to score better in visual-spatial skills testing than girls, while girls tend to outperform boys in verbal ability tests; both are important for good performance in STEM subjects, although the differences in the scores' distributions for the two genders are not large. However, they also show that there is more variability in boys' performance in quantitative and visual-spatial abilities compared to girls. This leads to a situation where there are relatively more boys than girls in the high-ability region of the test scores for visual-spatial and quantitative skills, despite the fact that the difference in the mean test scores are not so big. So, although there tend to be relatively more highly-able boys than girls, this does not mean that boys are significantly better at the skills traditionally associated with science than girls.

The paper also looks at evolutionary factors and differences in brain structures, both of which may have an impact. However, the paper makes it clear that a wide range of social and cultural influences are also important. The most important point they make for our purposes is that it is almost impossible to disentangle differences that arise from innate ability and external factors. Therefore, the bottom line is that, while biological factors may have an influence, there are many other factors that are as – if not more – important in influencing girls with regard to studying STEM subjects.

So it is clear that the lack of girls studying physics cannot simply be dismissed as being down to girls not being as good at it as boys. Good. I never believed that anyway! That being the case, the next question we must address is: what are the other factors that influence girls when it comes to deciding whether or not to study physics?

There has been extensive research into what these factors may be, with the evidence suggesting that families, teachers and schools, and wider society all play a significant role in influencing girls' attitudes towards science.

Family influence

There is no doubt that our parents have a substantial influence over all aspects of our lives. But how do they influence girls' attitudes towards science? The literature confirms that parents do have a major influence over subject choices (Lyons 2006), and shows that they also tend to have different academic and career expectations depending on the gender of their child (Gilmartin *et al.* 2006). There is also evidence to show that parents tend to believe that daughters are less interested in science than sons (Tenenbaum and Leaper 2003), and that science is harder for daughters than for sons. There is a correlation between a parent's belief in their child's interest and ability, and the child's actual interest and ability, particularly in the case of the beliefs of the mother. Fathers are also more likely to encourage gender-typed behaviour; so, for example, perceiving physics as a more 'male' subject, they will tend to be more active in encouraging sons to engage with it than daughters.

These findings are highlighted by a survey carried out by ICM on behalf of the Royal Society in 2010, which showed that only 18 per cent of respondents chose science as the best career choice for their (real or imagined) daughter, although, interestingly, more men than women selected 'scientist' as their top career choice.

Science is generally seen as difficult, both by children and by their parents; this is particularly true for the physical sciences (Lyons 2006; Williams *et al.* 2003).

I have a very clear memory of attending my school prize giving at which I received an award for academic achievement. Each award recipient was given a short introduction about what they had gone on to do with their hard-won A-levels. When it came to my turn, the audience were informed that I had gone to a good university to study physics – I still remember the ripple that went through the audience upon hearing my introduction. At the time I was very proud, but looking back now, I can see that it said more about the audience than it did about me. I was unusual because I was a woman and I was studying physics.

This impression of physics being very difficult is further exacerbated in families that have little 'science capital' (Lyons 2006; Archer *et al.* 2012), that is, those that do not have some sort of familial connection with science or scientists. Such families also tend to have a lack of knowledge about the wide range of career possibilities that are open to anyone with STEM qualifications.

Teacher/school influence

It is generally recognised that there is a loss in interest in the physical sciences that starts once pupils – both boys and girls – move to secondary school level (Osborne *et al.* 2003). Indeed, it appears that the culture of school science is the dominant factor in whether pupils choose to study the physical sciences (Lyons 2006).

It has been shown that boys are far more likely to report liking science than girls during their secondary school years, even though the loss in interest is observed for both genders (Osborne *et al.* 2003). However, it has been noted that simply liking science and/or being good at it is not enough: Pupils who would be regarded as falling into this category do not necessarily perceive science as a legitimate career aspiration (Aschbacher *et al.* 2010). One interesting but potentially contradictory point that is raised is that it appears that the secondary school years are when most practising female physicists believe that they first became attracted to the subject (Hazari *et al.* 2010). From my own perspective, that was certainly the case.

In order to understand how to counteract the decline in interest in the physical sciences among girls, it is important to find out what the contributory school- and teaching-related factors are. One way to understand this may be to compare girls' attitudes towards biology versus physics, as the level of interest in biology among students is known to remain good throughout secondary school, compared to the decline observed in the case of physics. A study has been made of differing attitudes towards biology and physics (Williams *et al.* 2003). It is clear that, regardless of gender, there is a correlation between the degree of difficulty and whether a student likes a subject or not; this is true whether considering biology or physics. What is particularly interesting in this study was that the students were asked to give reasons why they found the subject interesting or boring. The responses of students who stated that they found physics boring were divided by gender. This division revealed several interesting points. Although roughly the same number of girls and boys said that physics was boring because it was difficult, significantly more girls than boys stated that it was because it was too easy. An encouraging sign, at least from the point of view of capability! Significantly more girls than boys stated that physics was boring either because of the content of the subject or because they felt that it was irrelevant to real life. The relevance of physics to everyday life was also noted as an important factor for girls in other, earlier studies (Woolnough 1994). Interestingly, in a study of how girls and boys understand physics (Stadler *et al.* 2000), it appears that girls don't tend to believe that they understand a concept in physics until they can put it into a broader, non-scientific context, in marked contrast to boys. This may indicate why relevancy is a more important factor for girls. The good news is that it appears that the kind of contexts into which physics can be placed to make physics more relevant for girls also work for boys, although the reverse does not always appear to be the case (Stadler *et al.* 2000).

In studies performed during the 1990s into the factors that influence students' choice of physics (or lack thereof) (Woolnough 1994), it was demonstrated that the students were more likely to be encouraged to study physics, chemistry or engineering at degree level: if they had been taught by good teachers whose training was in the specific subject that they were subsequently teaching; if the subject was well structured, intellectually stimulating and relevant; and if they had had the opportunity to participate in extra-curricular activities relating to science (visits and talks, after-school clubs etc.).

The quality of the teaching is not the only factor directly relating to the teachers themselves. The student–teacher relationship is also an important factor. Speaking from personal experience, probably the deciding factor in my choice of career path was the positive and supportive relationship I had with my A-level physics teachers (one male, one female). From informal discussions with both male and female colleagues throughout my career, I know I am not alone in this. However, it is clear from the literature (Kenway and Gough 1998; Labudde *et al.* 2000; Brickhouse *et al.* 2000; Häussler and Hoffmann 2002) that negative experiences also impact on students' – particularly girls' – attitudes towards science. This can range from obvious lack of sympathy towards girls in science and bias in favour of male students in class, to more subtle effects that teachers themselves are probably not even directly aware of. In the latter case, this may be a lack of awareness of the effects in their own behaviour or of the internal dynamics of the group of students they are teaching (Guzzetti and Williams 1996). Work on the more subtle effects, at least in terms of increasing awareness among teachers, shows that this can be addressed successfully to everyone's benefit (Labudde *et al.* 2000; Guzzetti and Williams 1996).

It appears that it is not only the teachers' attitudes that need to be considered. Research has shown (Guzzetti and Williams 1996; Stake 2003) that the attitudes of male peers towards their female counterparts within the class is also an issue, and it has been postulated that this may have a negative impact on girls' performance and hence on the likelihood that they will continue to study science.

There are a number of issues relating to the delivery of the physics curriculum that are highlighted in the literature. For example, it has been observed that in many physics textbooks, if people are shown in the figures, they tend to be men (Guzzetti and Williams 1996). Flicking through my own A-level physics textbook now, I can see that this is certainly true, although I am not sure I was consciously aware of it at the time. To a certain extent, I believe that this is understandable, if not excusable: when the figure in question is a photograph, the fact that there are relatively few women in physics will lead to it being harder to find photographs of female physicists to include. Of course, this does not mean that authors and publishers should not try harder to find them. Even when the figure does not explicitly include a person, there is also the tendency to pick examples that are more meaningful to boys than girls (Guzzetti and Williams 1996).

Although it is well known that boys tend to dominate the discussions in physics classes, it has been shown that this may be down to the way in which the teacher runs the discussions. For example, it has been observed that girls respond better to different types of question style compared to boys (Stadler *et al.* 2000). The same study also observed that girls take longer to start using technical language. It is also well known that boys and girls tend to take on certain roles within practical work, with the boys tending to take on active roles that also allow them to dominate in this area (Stadler *et al.* 2000). Interestingly, it has been found that, amongst a mixed group of students who all claimed to find physics interesting, significantly fewer girls cited practical work as the main reason for the interest, compared to boys (Williams *et al.* 2003). Perhaps there is a correlation between these two observations?

Studies presented in the literature have shown that students tend to associate the traditional 'chalk-and-talk pedagogy' to learning with the teaching of the physical sciences, although most are critical of this approach and appreciate the efforts of teachers who attempt to make the subject more engaging (Lyons 2006). This style of teaching is believed to be more beneficial for boys than for girls (Arnot and Phipps 2003), suggesting that this may also be a contributory factor in girls' engagement with, and enjoyment of, physics.

In addition to the teaching of physics itself, it has been suggested that the curriculum lacks explicit teaching about STEM-related career opportunities, and that students of all ages lack awareness of the breadth of job opportunities that would be available to them with science qualifications (Archer *et al.* n.d.; Cleaves 2005). For example, a relatively recent study (Masnick *et al.* 2010) demonstrated that both boys and girls tend to perceive science professions as not being people-oriented and lacking creativity. The authors of the study suggest a number of ways in which these common misconceptions could lead to girls choosing STEM-related careers less than boys; for example, if the social dimension of their chosen profession is more important to women than men (which it certainly appears to be (Hazari *et al.* 2010)), then the perceptions of STEM-related professions formed at school may well lead girls away from choosing them as an adult. The authors of another study (Scherz and Oren 2006) note that it is not just children that possess stereotypical images of scientists; adults, even including teachers, do as well; in the case of teachers this can result in a negative effect on the way they teach science. However, they also conclude that teaching has very little effect on the images of scientists held by children; the media appear to be the main influence in this regard. Ofsted produced a report into girls' career aspirations that supports the findings of these studies, in that it found that from an early age, girls held stereotyped views about jobs for men and women (Ofsted 2011). It also concluded that careers education in schools did not focus enough on the skills needed by girls to handle issues such as career breaks or becoming a parent. It did note that a small number of the girls surveyed had their career ambitions changed through observations of and/or encounters with professionals in fields they had not previously considered. A report commissioned by WISE in 2012 observes that demonstrating the range of jobs available within STEM appears to be one of the most effective ways of attracting girls into STEM-related careers (WISE 2012).

Although I have considered family and school influences separately, it is important to remember that there will be some interplay between the two. Lyons (2006) considered this within the context of his study, observing that the physical sciences are more likely to be selected by students of either gender when both school and familial influences emphasise the subject's importance or strategic value. The negative impact of the perceived difficulty of the physical sciences also appears to be at least partially offset by strong familial support and good family relationships with the 'science advocate' within the family.

Wider societal influences

As noted in the above discussion of careers education, boys and girls tend to hold stereotyped views from an early age of what a scientist looks like and does. This image is predominantly of a middle-aged (or older) man who is bespectacled, wearing a lab coat, and is either bald or has wild, unkempt hair (Scherz and Oren 2006) – the classic Einstein-in-a-lab-coat look. Not only is the image stereotyped, it is generally a negative stereotype: scientists are overwhelmingly perceived as 'nerdy' (Buck *et al.* 2002). This stereotype has a particularly negative effect on girls, putting them off engaging with science education and impacting on their career choices (Buck *et al.* 2002). After all, who would want to become a scientist if it meant their future self was to be considered nerdy?

The image of a scientist is primarily formed and maintained through the media, suggesting that it is not just children who hold these views. Although there has been some success in recent years in improving the representation of scientists in the media ('the Brian Cox effect', for example), there is still a substantial under-representation of female scientists on TV, radio and in the newspapers.

A project was carried out in the early 2000s in which female scientists were taken into one class in an elementary school in the United States to see if this would help to overturn young children's stereotyped views of scientists (Buck *et al.* 2002). Interestingly, the children who participated in the study were not able to grasp that they were scientists and tended to assume that they must be teachers, that is, the presence of female scientists interacting with the children did not confront – let alone change – the stereotyped views held by the children. Another study (Hazari *et al.* 2010), which involved US college and university students and their high school physics experiences, also observed no significant influence on their likelihood to continue studying physics after high school from interactions with female scientists, such as visits by female guest speakers or discussions about the work carried out by female scientists. However, they did observe a positive impact on female students if they had been involved in a discussion of the under-representation of women in science while in high school; the discussion had no impact, positive or negative, on the male students.

A number of studies talk about 'self-efficacy' (e.g. Lyons 2006; Hazari *et al.* 2010), that is, a student's confidence in being successful in physics or science class. There is a direct connection between self-efficacy and achievement (Blin-Stoyle 1983), but there is a tendency for girls to have lower self-efficacy than boys. In a joint report from the Royal Society and the Institute of Physics in the early 1980s (Blin-Stoyle 1983), it was suggested that girls' less favourable attitude towards science (compared to boys) was related to the 'feminine role in society' and the fact that science has a generally masculine image. Physics was particularly singled out in this regard, being described in the report as 'the most remote of all the basic sciences from concern with living beings'.

In research conducted in the early 1990s (Williams *et al.* 2003), it was suggested that part of the reason for the relative popularity of biology versus physics could be related to the fact that the impression then was that the most exciting and challenging opportunities in science at the time appeared to be in biology-related areas. It will be interesting to see if, over the next few years, the impact of the switch-on of the Large Hadron Collider at CERN and the subsequent discovery of a 'Higgs-like' boson, along with other projects such as the Mars exploration rovers, will help to swing things back in favour of physics. It has certainly had a positive influence on the numbers of both male and female students looking to study doctorates in particle physics!

There is clearly a need to challenge stereotypes not just within schools but also in society as a whole, and the next section contains a few suggestions that could be considered as starting points to address this and the other issues that have been raised here. However, before discussing those, there is one further point that should be made, which is described and discussed well by Brickhouse *et al.* (2000). Most of the literature referenced here tends to make general statements about girls and boys, creating stereotypes of both genders that don't necessarily work in reality. It is therefore important to remember that the paths that lead both girls and boys into science are as unique as the individuals who follow them, and that we must not lose sight of this fact when attempting to encourage more girls into physics.

What can we do to improve the number of girls pursuing physics?

A number of the studies mentioned above have involved a direct intervention, in terms of teaching specially-designed lessons for a short period or the introduction of external speakers etc. (Hazari *et al.* 2010; Stadler *et al.* 2000; Labudde *et al.* 2000; Häussler and Hoffmann 2002; Stake 2003; Scherz and Oren 2006; Buck *et al.* 2002). Several of the studies make the point that any changes that are introduced to improve girls' engagement and achievements in physics should at least not be to the detriment of the boys in the class and, more generally, that a lot of the changes proposed ought to and do benefit boys and girls alike. For example, as already mentioned, when looking for the means to put physics into context in ways that will benefit girls, it has been noted that most of the contexts that have been reported to work for girls also work for boys (Stadler *et al.* 2000).

However, it would be reasonable to ask whether there is anything that teachers can do, at least at secondary school level, given that stereotypes of scientists are already well entrenched even at primary school level (Buck *et al.* 2002) and that the gender-stereotyping of physics also appears to begin during the early school years (Hazari *et al.* 2010). Well, the answer would appear to be yes, as it is also reported that many women who are now physicists first became interested in the subject during their high school years (Hazari *et al.* 2010). So there is clearly a window of opportunity to re-engage some girls who have already begun to lose their interest and desire to continue studying physics.

Another area in which improvements could be made is in the use of more gender-neutral examples in the classroom, and the use of a better mix of images in terms of portraying images of male and female scientists in physics textbooks (Guzzetti and Williams 1996). It has also been recommended that teachers should receive training to sensitise them to the differing needs of boys and girls in the science classroom, as this was found to have a beneficial effect for both students and teachers (Labudde *et al.* 2000).

In the study that compared students' attitudes to biology versus physics (Williams *et al.* 2003), it is suggested that a more interdisciplinary approach could be taken. For example, given the greater popularity of biology compared to physics, efforts could be made to emphasise places where physics is important when teaching the more popular areas of biology. It is also suggested that when teaching the physics curriculum, more should be made of the individual topics that students find more interesting in order to maintain their interest in the subject as a whole. Obviously we can not just ignore the parts of the curriculum that are considered to be boring by students, but extra activities relating to the interesting parts may be beneficial.

As already discussed, *how* the material is taught is as important as the choice of topics (Lyons 2006). Students, both boys and girls alike, have been reported to be critical of the traditional passive-receiver, teacher-centred approach to delivery of science lessons. Making the effort to engage students more actively in the learning process would address this point and have the added benefit of engaging more of the female students.

A number of studies have addressed the issue of challenging stereotypes of scientists and science. There is evidence to suggest that discussing the issues surrounding the under-representation of women in physics has a positive impact on girls and should therefore be undertaken (Hazari *et al.* 2010). However, the conclusions are less clear-cut in the case of the introduction of female role models into the classroom. It is clear that steps should be taken to confront the stereotypes held by both boys and girls, but it is less clear that, in order to encourage the latter, female scientists are specifically required to do this or are most effective. The studies described above report no beneficial effect from the use of female scientists; however, a report commissioned by WISE in 2012 reports that a Girl Guiding UK survey conducted in 2011 found that 60 per cent of respondents were put off by the lack of female role models in science (WISE 2012).

Conclusion

In conclusion, the different influences that impact on girls' decisions as to whether to study, or continue studying, physics are many and varied. It also seems that is it pretty well impossible to pick out one or two key factors that, if addressed appropriately, would solve the problem. This is clearly true from the fact that this issue has been discussed in various forms for at least the last thirty years – had there been a simple solution, one might expect it to have been adopted with by now!

It is clear, however, that there are two main strands that can and should be addressed in any serious attempt to solve this problem: The influence of schools and teachers on girls' attitudes towards physics and that of the society in which we live. Whilst the latter is something that teachers can have only a relatively limited impact on, the former is clearly something that every physics teacher should have in mind while teaching the subject. There are clearly plenty of things that can be done within schools to improve the engagement of girls with physics. I wish you the best of luck!

Key questions

1   What is your own mental image of a physicist, and what impact do you think it has on your approach to teaching the subject?

2   Are there any aspects of your teaching, either in style or content, that could be unintentionally favouring male students, and what could you do to modify them to make them more inclusive?
3   Are there other things that could be done within schools to help address this issue, such as events, visits, after-school clubs etc.?

Further reading

Murphy, P. and Whitelegg, E. (2006) *Girls in the physics classroom: A review of the research on the participation of girls in physics.* Institute of Physics report, June. ('Red Book 1'.) Recommended for more detail on the research that has been done into girls and physics.

Daly, A., Grant, L. and Bultitude, K. (2009) *Girls into physics – action research.* Report DCSF-RR103. Available online at https://www.education.gov.uk/publications/standard/publicationDetail/Page1/DCSF-RR103. Likewise recommended for more detail on the research into girls and physics.

Institute of Physics (2006) *Girls in the physics classroom: A teacher's guide for action.* Institute of Physics report, December. ('Red Book 2'.) Provides guidance and ideas about how you can address some of the issues that have been raised here and elsewhere in the classroom.

Institute of Physics (n.d.) *It's different for girls: How can senior leaders in schools support the take-up of A-level physics by girls?* Briefing sheet. Available online at: http://www.iop.org/girlsinphysics.

Institute of Physics (n.d.) *It's different for girls: The influence of schools.* Briefing sheet. Available online at: http://www.iop.org/girlsinphysics. The Institute of Physics provides some excellent briefing sheets that may be helpful if you wish to raise awareness within your school community about the issues relating to girls' perceptions of physics. This sheet and the following one are examples.

ASPIRES project (n.d.) ASPIRES: Science aspirations and career choice: Age 10–14. King's College London. Available online at: http://www.kcl.ac.uk/sspp/departments/education/research/aspires/index.aspx. ASPIRES is an ongoing research project into the more general uptake of science amongst 10–14 year-olds. The website is full of useful information, allowing you to keep up with issues in this area, and also providing the opportunity to get involved in their research if you wish to.

References

Archer, L., DeWitt, J., Osborne, J., Dillon, J., Willis, B., and Wong, B. (2012) Science aspirations and family habitus: How families shape children's engagement and identification with science. *American Educational Research Journal*, Online-First, January: 1–28.

Archer, L., Osborne, A., and DeWitt, J. (n.d.) *Ten science facts and fictions: The case for early education about STEM careers.* ASPIRES project, King's College, London. Available online at: http://www.kcl.ac.uk/sspp/departments/education/research/aspires/index.aspx (25 June 2013).

Arnot, M., and Phipps, A. (2003) *Gender and education in the United Kingdom.* UNESCO. Available online at: http://unesdoc.unesco.org/images/0014/001467/146735e.pdf (25 June 2013).

Aschbacher, P., Li, E., and Roth, E. J. (2010) Is science me? High school students' identities, participation and aspirations in science, engineering and medicine. *Journal of Research in Science Teaching*, 47(5): 564–82.

Blin-Stoyle, R. (1983) Girls and physics. *Physics Education*, 18: 225–8.

Brickhouse, N. W., Lowery, P., and Schultz, K. (2000) What kind of a girl does science? The construction of school science identities. *Journal of Research in Science Teaching*, 37(5): 441–58.

Buck, G. A., Lesley-Lelecky, D., and Kirby, S. K. (2002) Bringing female scientists into the elementary classroom: Confronting the strength of elementary students' stereotypical images of scientists. *Journal of Elementary Science Education*, 14(2): 1–10.

CBI (2011) Building for growth: Business priorities for education and skills, education and skills survey. Available online at: http://www.cbi.org.uk/business-issues/education-and-skills/in-focus/education-and-skills-survey/ (25 June 2013).

Cleaves, A. (2005) The formation of science choices in secondary school. *International Journal of Science Education*, 27(4): 471–86.

Gilmartin, S., Li, E., and Aschbacher, P. (2006) The relationship between interest in physical science/engineering, science class experiences, and family contexts: Variations by gender and race/ethnicity amongst secondary students. *Journal of Women and Minorities in Science and Engineering*, 12: 179–207.

Guzzetti, B. J., and Williams, W. O. (1996) Gender, text and discussion: Examining intellectual safety in the science classroom. *Journal of Research in Science Teaching*, 33(1): 5–20.

Halpern, D. F., Benbow, C. P., Geary, D. C., Gur, R. C., Shibley Hyde, J., and Gernsbacher, M. A. (2007) The science of sex-differences in science and mathematics. *Psychological Science in the Public Interest*, 8: 1–52.

Häussler, P., and Hoffmann, L. (2002) An intervention study to enhance girls' interest, self-concept and achievement in physics class. *Journal of Research in Science Teaching*, 39(9): 870–88.

Hazari, Z., Sonnert, G., Sadler, P. M. and Shanahan, M.-C. (2010) Connecting high school physics experiences, outcome expectations, physics identity and physics career choice: A gender study. *Journal of Research in Science Teaching*, 47(8): 978–1003.

Kenway, J., and Gough, A. (1998) Gender and science education in schools: A review 'with attitude'. *Studies in Science Education*, 31(1): 1–29.

Kirkup, G., Zaleyski, A., Maruyama, T., and Batool, I. (2010) *Women and men in science, engineering and technology: The UK statistics guide 2010.* Bradford: UKRC.

Labudde, P., Herzog, W., Neuenschwander, M. P., Violi, E., and Gerber, C. (2000) Girls and physics: Teaching and learning strategies tested by classroom interventions in grade 11. *International Journal of Science Education*, 22(2): 143–57.

Lyons, T. (2006) The puzzle of falling enrolments in physics and chemistry courses: Putting some pieces together. *Research in Science Education*, 36: 285–311.

Masnick, A. M., Valenti, S. S., Cox, B. D., and Osman, C. J. (2010) A multidimensional scaling analysis of students' attitudes to science careers. *International Journal of Science Education*, 32(5): 653–67.

Ofsted (2011) *Girls' career aspirations* (April). Available online at: http://www.ofsted.gov.uk/publications/090239 (25 June 2013).

Osborne, J., Simon, S. and Collins, S., (2003) Attitudes towards science: A review of the literature and its implications. *International Journal of Science Education*, 25(9): 1049–79.

Scherz, Z., and Oren, M. (2006) How to change students' images of science and technology. *Science Education*, 90(6): 965–85.

Stadler, H., Duit, R., and Benke, G., (2000) Do boys and girls understand physics differently? *Physics Education*, 35(6): 417.

Stake, J. E. (2003) Understanding male bias against girls and women in science. *Journal of Applied Social Psychology*, 33(4): 667–82.

Tenenbaum, H. R. and Leaper, C. (2003) Parent–child conversations about science: The socialization of gender inequities? *Developmental Psychology*, 39(1): 34–47.

UKRC (2010) *Progress Newsletter*, issue 22 (August). Bradford: UKRC.

Williams, C., Stanisstreet, M., Spall, K., Boyes, E., and Dickson, D. (2003) Why aren't secondary students interested in physics? *Physics Education*, 38(4): 324.

WISE (Women into Science and Engineering) (2012) *Engaging girls in science, technology, engineering and maths: What works?* Summary findings from a research review for WISE, sponsored by BAE Systems, July. Available online at: http://www.wisecampaign.org.uk/files/useruploads/files/resources/wise_report_july_2012_bae_systems_what_works_summary.pdf (accessed 25 June 2013).

Woolnough, B. E. (1994). Why students choose physics, or reject it. *Physics Education*, 29: 368.

Chapter 5

Improving primary to secondary transition: 'now let's teach them properly'

Michael Allen

Introduction

With contemporary media reports teeming with stories relating to topics such as climate change, DNA profiling, parachute jumps from the edge of space and the possibility of extraterrestrial life on the moons of Jupiter, scientific literacy has become increasingly important in everyday life. However, interest in science at secondary school level is on the decline. Since children tend to have a keen interest in science during their years at primary school, blame has been placed squarely on the shoulders of curriculum designers, secondary schools and science teachers themselves. Something appears to be lost after the transfer from primary to secondary school that is never regained. This chapter aims to summarise a selection of recent research on primary to secondary transfer, bringing a critical eye to this work with a view to suggesting new ways forward that may help alleviate these problems.

Part 1: outlining the problems

Dips in interest

Studies that have focused on the transfer of pupils from primary to secondary school in the United Kingdom have invariably reported that, post-transfer, pupils become less interested in science. This dip in attitude is not geographically limited to the United Kingdom, but also occurs overseas, particularly in more developed countries (Martin *et al.* 2008). The same children who experience a loss of interest in science have either less or no decline in attitudes towards English (Galton *et al.* 1999). Even well after transfer, pupils' attitudes towards science lessons continue to dip from Year 7 to Year 9. Although studies have found a deterioration in attitudes with English and mathematics as well over this period, there is a greater dip with science (Barmby *et al.* 2008). Several studies have shown that girls experience a greater decline in their attitude to science after transfer (e.g. Francis and Greer 1999) and beyond (e.g. Hargreaves and Galton 2002). In addition to the dip in interest after transfer, comparing the TIMSS data between 1995 and 2007, it can be concluded that there has been an overall fall with respect to positive attitudes to science at both primary and secondary ages (Diack 2009).[1]

These are worrying phenomena that do not bode well for the scientific literacy of the populace, at a time when there are increasing demands on the citizen to be able to understand scientific ideas reported in the media. A number of possible reasons for these declines in interest have been offered by writers and researchers, and fall into two general categories.

I   Repetition of work

Despite an expectation of learning novel and exciting science, pupils new to Year 7 are frequently disappointed. They find that secondary science lessons cover familiar topics that focus on the same concepts that they had to learn during their primary years. They quickly become bored, then eventually disillusioned with science, and there is some evidence to suggest that the brighter pupils are affected the most (Pell *et al.* 2007).

Decisions to repeat Key Stage 2 content at Key Stage 3 have been based on sound theoretical reasoning. Bruner's idea of a 'spiral curriculum', which lies at the heart of many science curricula, involves revisiting the same topic at different times during schooling with each subsequent visitation dealing with ever more sophisticated concepts (Bruner 1996). In addition, enabling a pupil to meet the same material repeatedly over time allows for enhanced learning through consolidation. That said, the science that new secondary pupils encounter is rarely at a more challenging level to that in primary school, and teachers use the same experiments and modes of delivery (Morrison 2000), which culminates in not a spiral but a circular curriculum.

Other reasons for repetition of content are based on more tenuous assumptions. A *tabula rasa* or clean slate approach taken by many secondary departments assumes that the science previously taught at primary school cannot be considered of value. This is due to either an ignorance of the primary science curriculum (Braund and Driver 2005) or a detrimental view of the general quality of primary science (Nott and Wellington 1999). Alternately, secondary teachers may hold the belief, often with some justification, that children from different primary feeder schools come with markedly differing levels of science understanding and knowledge, so one must begin teaching with the lowest level in mind. Approaches that have attempted to solve the latter problem of a variance of previous science experience (e.g. bridging units) are discussed later in the chapter.

Galton (2009) has noted that science in particular suffers compared with other subjects in the curriculum that are less bound to a spiral curricular design. For instance, history and geography curricula have self-contained topic areas, such as the Ancient Egyptians or the Tropical Rainforest, that, once taught, are seldom revisited later during a child's schooling. In contrast, key scientific ideas such as those inherent in, say, a forces topic are introduced in Key Stage 1, and then reappear several times all the way up to the end of Key Stage 4. Such a spiral curricular approach can just seem to pupils who are keen to learn novel, stimulating ideas like a pointless going over of ground already covered.

2 Pedagogical style

As pupils arrive at secondary school they have high expectations about the science that they are about to be taught. They look forward to the new experience of working 'scientifically' in sophisticated laboratories that are purpose-built (Galton 2009), and it is well established that pupils find practical work motivating (e.g. Hodson 1993). However, the true nature of secondary science – that of a very crowded curriculum – means that written work takes up the lion's share of lesson time, and when practical work is carried out this is accompanied by further writing tasks such as the tedious writing up of experiments (Galton 2009). These reasons have been cited by pupils complaining why science is so unpopular (Hargreaves and Galton 2002). Further compounding the problem, traditionally, primary school pedagogy centres around small groups where children work together in a collaborative way, and many pupils find it difficult to adjust to the change to more teacher-led, didactic pedagogies on transfer (Moore 2008). The more didactic pedagogies practiced by secondary teachers may be a result of a need to manage pupil behaviour, which can deteriorate once pupils feel established in secondary school.

That said, there is increasing evidence to suggest that Year 6 pupils are spending significant amounts of time in primary schools working towards their performance in the end of Key Stage 2 National Tests in English and Mathematics and, as a result, the teaching in Year 6 is becoming more didactic. Galton (2007) investigated teacher statements in the primary classroom and found that since 1976 statements of fact had doubled. Tables are more likely to be organised in rows (Webb and Vulliamy 2006), and teaching is interrogative and directive (Smith *et al.* 2004). These factors point to the prevalence of 'teaching to the test' pedagogies in preparation for the National Tests; for instance, whole-class revision periods are becoming more common (Galton 2009). Thus, pupils who expect on transfer that the change in school will bring about different, more engaging forms of teaching are disappointed when they find similar pedagogies at work in Year 7 science classrooms to those they have experienced in Year 6 (Hargreaves and Galton 2002).

Dips in attainment

Apart from a deterioration in pupils' attitudes towards science after transfer, there is strong evidence that their rate of learning decreases, or in some cases their knowledge and understanding of scientific concepts even degenerates to a level that is lower than when they left primary school. Of course, any dip in *attitude* may be causative for such a dip in attainment.

Pupils commonly experience a period of adjustment during the period immediately after transfer that is related to a drop in performance, though for most pupils this is transient and they have usually settled in by the end of the first term (Galton 2009). However, a minority suffers dips in attainment that are more sustained: Chedzoy and Burden (2005) have estimated that this may affect 10 per cent of pupils. More seriously, some pupils lose ground with respect to their science learning. End of Key Stage National Test data have indicated a drop in science levels for a significant number of pupils over the three year period from the end of Year 6 to the end of Year 9; this does not occur in English and mathematics (Braund 2009). In his examination of End of Key Stage target levels, Braund (2008) found that fewer pupils hit their science target level in Year 9 compared with Year 6, suggesting that their rate of learning in science lessons has regressed – over a given time they are making far less progress with their secondary science than they did with primary science. Although English and mathematics experience a slight regression, this is nowhere near as high as with science.

An awareness of a dip in science attainment between Key Stages 2 and 3 is nothing new. In the mid 1990s the plaudits of primary science were much hailed, and secondary science departments were berated for not continuing this good work, signalled by the fact that their pupils had not demonstrated satisfactory progress since they had left primary school (Ofsted 2000). Possible reasons that have been suggested for the dip in science attainment on transfer are discussed next, and can be categorised into three broad groups.

I Distrust of primary schools' assessed levels

Related to the previously-discussed view held by some secondary science teachers, that primary schools teach children a somehow inferior form of science, there can also be a distrust of the assessed levels to which primary teachers have allocated pupils at the end of Year 6 (Nott and Wellington 1999). This belief can be politically-driven, based on an assumption that primary schools need to show that their pupils have made as much progress as possible by the end of Year 6. It is exacerbated by a need for secondary schools to similarly show that pupils have progressed with their learning, and so it is in those schools' interests to record as low a level as possible for baseline scores at the start of Year 7. Until recently there were Key Stage 2 National Tests in Science, the 'Year 6 science SATs', and so assessment at the end of primary schooling could be argued to have had an acceptable amount of rigor that secondary teachers could not challenge. Since these tests were scrapped after 2009, primary teachers have had to use more subjective measures to assess pupils' science attainment in Year 6, which can only serve to further fuel these fires of criticism. Interestingly, it appears that secondary teachers have more confidence in levels of practical work that have been assessed by primary schools than in levels of substantive science knowledge (Peacock 1999).

As a consequence of this, secondary science teachers can underestimate pupils' knowledge and understanding on transfer. This includes how familiar Year 7 pupils are with scientific language and vocabulary. When primary and secondary teachers were asked about the science terminology and concepts they expected their pupils to know, there was a notable decrease in the numbers and complexity of words from Key Stage 2 to 3 (Peacock 1999). The fact that primary teachers expected their children to know and understand more science than secondary teachers did is surprising, but indicative of an underestimation of the generally high standard of primary science that results in an all too-common *tabula rasa* stance and subsequent repetition of content at Key Stage 3.

2   Lack of a continuity of learning

Secondary science teachers have tended to have a general ignorance of the content of primary science curricula (Braund and Driver 2005). The National Curriculum Programme of Study was designed so that there is a smooth transition between Year 6 and 7 science teaching, although this is frequently not the case. Secondary teachers may have little knowledge about the science that pupils have learned in primary school, which leads to an uncoordinated shift at transition where, for example, topics are repeated.

Galton *et al.* (2003) noted that prior to the turn of the twenty-first century most research into transfer focused on the social and emotional aspects of the pupil experience, for instance how they adjusted to the change in school, and did not include comments on the continuity and quality of learning between Key Stages 2 and 3. However, more recently, studies have looked at academic and pedagogical issues at transfer with a view for a more co-ordinated transition, and advise more communication between primary and secondary schools as a means to solving the issue (ibid.).

A further reason for the attainment dip at transfer could be the quality of science teaching in some secondary schools (Hayes and Clay 2004). Secondary science has been a shortage subject for a number of years, and many schools find it very difficult to recruit well qualified science teachers, physics specialists particularly. This has resulted in science departments employing unqualified teachers, often overseas trained, on short term contracts in order to fill gaps. In many cases, pupils have to endure a constant stream of different supply teachers, which can only create disjointed science teaching and learning, and a loss of continuity. For subjects such as English, there is less of a labour shortfall, and this may partly explain dips in attitude and attainment that are specific to science. Only when all schools are able recruit appropriately-qualified, enthusiastic science teachers will this issue begin to be addressed.

3   Secondary pedagogy

For reasons previously discussed, on transfer, pupils can be disappointed about the nature of secondary science teaching because it does not live up to what was expected. This leads them to switch off from science and, as their attitudes deteriorate, achievement levels can fall.

Part 2: solving the problems

The remainder of the chapter offers an account of the strategies and practices that schools and teachers could adopt to help alleviate the issues of declines in interest and attainment at the point of transfer between Key Stages 2 and 3, and beyond.

If science becomes understandable, it then becomes interesting

The reasons for a loss of enthusiasm about science may be linked to the more complex scientific ideas presented by secondary teachers. As the spiral science National Curriculum continues throughout the years of Key Stages 3 and 4, ideas do become more complex and many pupils are unable to fully grasp them. These individuals then become de-motivated, permanently switching off from science that they view as being 'too hard'. This would help to explain that dips in motivation are higher when science is compared with English, a less directive subject that allows pupils to be more creative, and which uses less complex, seemingly-ungraspable ideas. Over the past fifteen years, the increased uptake of subjects at A-level such as media studies and drama, alongside the demise of physics and chemistry, is in part due to their perceived 'softness' and lack of complexity (Jenkins and Pell 2006).

Once secondary teachers lose the interest of pupils in this way it is very difficult to re-motivate them. The mental effort required to understand many science concepts demands that time and hard work be afforded during scientific study if these concepts are to be grasped, and adequate effort is never going to be given by a pupil who has become disinterested. Thus, any success with learning that may actually cause the individual to once more become enthusiastic about science is impeded by a barrier of perceived difficulty. A draft primary National Curriculum Programme of Study recently published (Department for Education 2012) lays out the probable future direction that primary science will take, and it appears it will be more hard-edged and academic than is currently the case. This can only exacerbate the situation, and will perhaps even cause a lessening of interest in science at the primary level because of the increased difficulty of the concepts.

The science education literature contains vast amounts of studies that suggest ways of making science concepts more understandable for pupils. Effective interventions often use strategies based on the tenets of the conceptual change paradigm (Allen 2010). The correction of science misconceptions at all levels is a continuing battle for primary teachers and secondary science specialists alike, and if effective pedagogies were used more often in the classroom then this would clearly go some way to making science appear understandable. The more pupils who enjoy success with their science learning, the less we will lose as a result of the perceived impenetrability of many science concepts.

Key Stage 3 and 4 teachers are called upon to deliver the more complex ideas, so there is a clear need to try new approaches to teach science effectively in a way that is understandable for all pupils. At present, innovative pedagogies that have been shown to be successful by science education researchers and are reported in the literature are seldom brought to the attention of qualified secondary specialists. One exception to this would be the interventions summarised in the Association for Science Education (ASE) journals *School Science Review* and *Primary Science*, but only a fraction of science teachers are ASE members so access to this material is limited. The recent decline of Local Education Authority advisory services means that there is an increasing lack of co-ordinated INSET in schools, which only serves to intensify this barrier to further professional development of pedagogic skill.

Bridging units

As previously discussed, secondary science departments have the tendency to produce *tabula rasa* schemes of work that do not consider the excellent science teaching carried out in primary schools. The reasons for this approach include a lack of awareness of the primary curriculum. Even when secondary teachers acknowledge that primary schools play a significant part in the science education of pupils, the fact that the children arrive at secondary school from different feeder schools creates the assumption that the variety of experiences warrants that everyone must start from scratch. This culminates in a break in the continuity of the curriculum and repetition of primary work, leading to declines in interest. One way to solve this problem would be for primary and secondary schools to co-ordinate with each other closely at the point of transfer. In some cases this has involved primary teachers delivering some material to Year 6 pupils that has been prepared in close conjunction with colleagues in secondary science departments, with a view to providing more of a seamless transition between Key Stage 2 and 3; such material has been called a bridging unit.

Galton *et al.* (2003) carried out an evaluation of QCA-developed bridging units and report a mixed reception from teachers. Primary teachers found that the introduction of extra work into an already overcrowded curriculum meant that non-core subjects were neglected as a result. The teachers also thought that the intensive science teaching that had been already carried out in Year 6 as preparation for the National Tests meant that children had already spent an inordinate amount of time learning science at the expense of other subjects, and to devote further time was neither necessary nor desirable. Secondary teachers also expressed some reservations with primary schools using bridging units, as not all feeder schools utilised them and so they would only have benefits for a proportion of pupils in Year 7. However, there were clear benefits with regard to content not being repeated, and a better understanding of Key Stage 2 work by secondary teachers.

Providing appropriate role models

It is well documented that, in the secondary years when pupils are considering their career choices, a lack of positive role models is a factor for the decline in interest in the physical sciences (Bevins *et al.* 2005). As well as science being perceived as overly difficult, scientists (along with mathematicians) are viewed as geeky and 'uncool'. It is socially acceptable therefore for teenagers not to be proficient at these subjects, and also to dislike them, because the consensus is that they are very hard and unpopular. Recent research has determined that children as young as ten years old are considering science as a career (e.g. Archer *et al.* 2012), and so primary teachers should be thinking about the quality of science role models that their pupils are presented with. Since this important issue appears to be extant during the transfer period between Key Stages 2 and 3, this further underlines the importance of having a co-ordinated transition.

It is beyond the scope of this chapter to discuss in detail the extensive literature that deals with role models in science; suffice it to say that there are several initiatives aimed at promoting a more positive and relevant role model to teenagers, for instance, assembly talks given by practicing scientists. This is particularly the case with the current drive to get more girls interested in the physical sciences (Murphy and Whitelegg 2006).

Conclusion and final recommendations

The recommendations suggested in this chapter would mean that serving teachers change their practices in order to ensure that the transfer from Key Stage 2 to 3 is as smooth as possible for pupils. There is strong research evidence to show that certain approaches work. However, when these approaches have been introduced or piloted they have been located in geographical enclaves, for instance, the work by the University of York with bridging units. Clearly, pupils who have been exposed to these approaches have benefited, but ideally changes need to be implemented on a national scale. Research has shown that it is often difficult to change teachers' behaviours and practices by introducing new innovations (Yair 2000). This chapter ends with three final recommendations on how the way teachers' teach could be changed on a long term basis, arguing that the best way forward would be the introduction of prescriptive governmentally-derived top-down measures.

The National Strategy initiatives were presented and administered in England and Wales by the then Department for Education and Skills (DfES) at the turn of the twenty-first century. Their introduction was heralded by intensive in-service training funded by DfES and delivered by consultants on a truly national scale.

The Strategy initiatives brought about wholesale changes in the way teachers taught their lessons in all state-maintained schools, particularly in the way literacy and numeracy was taught at the primary level. Schools adopted the 'three-part lesson', an approach that has remained viable today. Some writers have noted that the Strategies were not entirely successful, for instance in promoting more collaborative group work during Key Stage 3 science lessons (Stoll *et al.* 2003). However, if a new, government-funded initiative involving compulsory INSET focused on the problems associated with transfer were implemented, similar permanent changes to teachers' behaviours and attitudes may ensue.

Secondly, the current Department for Education (DfE) has powers to lay down specific requirements that schools must adhere to in relation to inspection matters, with these requirements later being examined and enforced by the Office for Standards in Education (Ofsted). If initiatives that are known to improve the quality of transfer, such as bridging units, are introduced into Ofsted criteria, then schools would be compelled to abide by them. Initiatives that help primary schools make closer links with secondary science departments would certainly help secondary teachers better understand the work that children carry out in Key Stage 2, thus avoiding *tabula rasa* effects and helping them appreciate the value of primary science, including standards of assessment.

Lastly, the introduction of the new primary and secondary National Curriculum Programmes of Study in 2014 provides a golden opportunity to advance the issue of transfer from Key Stages 2 to 3. If statutory tasks were to be embedded in the Programme of Study, then the manner in which both primary and secondary schools deal with transfer could be influenced in a legislative way. Importantly, there is scope to direct how science is taught at Key Stage 3 in order that it lives up to the high expectations of Year 7 pupils, and is also more understandable.

Key questions

1   What are the factors that would hinder the implementation of a national programme of professional development for secondary science teachers that was focused on more effective pedagogy?
2   From 2015 all young people in England aged 16–18 years will be required to remain in education or training. How could post-16 science curricula meaningfully engage with these students, so laying the foundations for the creation of a populace that has good scientific literacy?

Further reading

Galton, M. (2009) Moving to secondary school: Initial encounters and their effects. In *Perspectives on Education 2: Primary–Secondary Transfer in Science.* London: The Welcome Trust (pp. 5–21). This article presents a comprehensive review that focuses on current perspectives, and contrasts the different views of three writers working in the field of primary to secondary transfer.

Note

1 A report on the most recent TIMMS data is due to be published in December 2012.

References

Allen, M. (2010) *Misconceptions in Primary Science*. Maidenhead: Open University Press.

Archer, L., DeWitt, J., Osborne, J,. Dillon, J., Willis, B., and Wong, B. (2012) Science aspirations, capital, and family habitus: How families shape children's engagement and identification with science. *American Educational Research Journal*, 49: 881–908.

Barmby, P., Kind, P., and Jones, K. (2008) Examining secondary school attitudes in secondary science. *International Journal of Science Education*, 30: 1075–93.

Bevins, S., Brodie, M., and Brodie, E. (2005) *A study of UK secondary school students' perceptions of science and engineering*. Paper presented at the European Educational Research Association Annual Conference, Dublin, 7–10 September.

Braund, M. (2008) *Starting Science...Again?* London: Sage.

Braund, M. (2009) Progression and continuity in learning science at transfer from primary to secondary school. In *Perspectives on Education 2: Primary–Secondary Transfer in Science*. London: The Welcome Trust (pp. 22–35).

Braund, M. and Driver, M. (2005) Pupils' perception of practical science in primary and secondary school: Implications for improving progression and continuity of learning. *Educational Research*, 47: 77–91.

Bruner, J. (1996) *The Process of Education*. Cambridge, MA: Harvard University Press.

Chedzoy, S. M., and Burden, R. L. (2005) Making the move: Assessing student attitudes to primary–secondary transfer. *Research in Education*, 74: 22–35.

Department for Education (2012) *National Curriculum for Science: Key Stages 1 and 2 – Draft*. London: Crown Copyright.

Diack, A. (2009) A smoother path: Managing the challenge of school transfer. In *Perspectives on Education 2: Primary–Secondary Transfer in Science*. London: The Welcome Trust (pp. 39–51).

Francis, L., and Greer, J. (1999) Measuring attitude towards science among secondary school students: The affective domain. *Research in Science and Technological Education*, 17: 219–26.

Galton, M. (2007) *Learning and Teaching in the Primary Classroom*. London: Sage.

Galton, M. (2009) Moving to secondary school: Initial encounters and their effects. In *Perspectives on Education 2: Primary–Secondary Transfer in Science*. London: The Welcome Trust (pp. 5–21).

Galton, M., Gray, J., and Rudduck, J. (1999) *The Impact of School Transitions and Transfers on Pupil Progress and Attainment*. London: Department for Education and Skills.

Galton, M., Gray, J., Rudduck, J., Berry, M., Demetriou, H., Edwards, J., Goalen, P., Hargreaves, L., Hussey, S., Pell, T., Schagen, I., and Charles, M. (2003) *Transfer and Transitions in the Middle Years of Schooling (7–14): Continuities and discontinuities in learning*. Research Report RR443. London: Department for Education and Skills.

Hargreaves, L., and Galton M. (2002) *Moving from the Primary Classroom: 20 years on*. London: Routledge.

Hayes, S. G., and Clay, J. (2004) *Progression from Key Stage 2 to 3: Understanding the context and nature of performance and underperformance at Key Stage 3*. Paper presented at the British Educational Research Association Annual Conference, University of Manchester, 16–18 September.

Hodson, D. (1993) Rethinking old ways: Towards a more critical approach to practical work in school science. *Studies in Science Education*, 22: 85–142.

Jenkins, E., and Pell, T. (2006) *The Relevance of Science Education Project (ROSE) in England: A summary of findings*. Leeds: Centre for Studies in Science and Mathematics Education, University of Leeds.

Martin, M., Mullis, I., and Foy, P. (2008) *TIMSS 2007: International Science Report*. Boston: TIMSS and PIRLS International Study Centre, Lynch, School of Education, Boston College.

Moore, C. (2008) Bridging the divide: Parts 1 and 2. In M. Braund (ed.) *Starting Science … Again?* London: Sage (pp. 63–98).

Morrison, I. (2000) School's great – apart from the lessons: Sustaining the excitement of learning post transfer. *Improving Schools*, 3: 46–9.

Murphy, P., and Whitelegg, E. (2006) *Girls in the Physics Classroom: A review of the research on the participation of girls in physics*. London: Institute of Physics.

Nott, M., and Wellington, J. (1999) The state we're in: Issues in Key Stage 3 and 4 science. *School Science Review*, 81: 13–8.

Ofsted (2000) *Progress in Key Stage 3 Science*. London: Ofsted.

Peacock, G. (1999) *Continuity and progression between key stages in science*. Paper presented to the conference of the British Educational Research Association, University of Sussex, 2–5 September.

Pell, T., Galton, M., Steward, S., Page, C., and Hargreaves, L. (2007) Group work at Key Stage 3: Solving an attitudinal crisis among young adolescents? *Research Papers in Education*, 22: 309–32.

Smith, F., Hardman, F., Wall, K., and Mroz, M. (2004) Interactive whole class teaching in the National Literacy and Numeracy Strategies. *British Educational Research Journal*, 30: 395–412.

Stoll, L., Stobart, G., Martin, S., Freeman, S., Freedman, E., Sammons, P., and Smees, R. (2003) *Preparing for Change: Evaluation of the implementation of the Key Stage 3 Pilot Strategy. Final Report*. London: Department for Education and Skills.

Webb, R., and Vulliamy, G. (2006) *Coming Full Circle? The impact of New Labour's education policies on primary school teachers' work*. London: Association of Teachers and Lecturers.

Yair, G. (2000) Educational battlefields in America: The tug of war over students' engagement with instruction. *Sociology of Education*, 73: 247–69.

Chapter 6

The secondary science curriculum: a grey and dreary landscape?

Rob Toplis

Introduction

> 'Today there's good news and there's bad news', announces the science teacher to a class of 15-year-olds. 'The good news is that we are now doing some chemistry.'
>
> 'What's the bad news?' asks a student on the second row.
>
> 'It's the blast furnace', replies the teacher. The class groans.

What is it about the mere thought of studying the blast furnace that results in an overwhelming lack of enthusiasm? For many, this is just one example of the grey and dreary landscape experienced by many students in school.

This chapter reviews the debates surrounding the science curriculum and how school science is taught in the secondary phase of education, ages 11 to 19 years. In doing so, it discusses issues around some fundamental questions such as: What is the problem? What is the current science education diet of students? What sort of science education might we want? Who do we want it for? How might this be achieved? In discussing these tensions in the science curriculum, the debate extends to international perspectives on: the uptake of science subjects; school students' attitudes to science; problems with the transition from the primary or elementary phase (ages 5 to 11 years) to the secondary phase (ages 11 to 16/19 years) of education; influences on the science curriculum; notions of scientific literacy; and what sort of science education is really needed.

What's the problem?

In one sense, science education is facing a crisis. An advanced technological society brings with it a global demand for people who are educated in scientific and engineering subjects. The response to this demand is to have a school science education that is capable of producing sufficient numbers of students at age 18 to 19 years who can be recruited into higher education or training in science, technology, engineering and mathematics (STEM) subjects. Traditionally, many of these students in the United Kingdom would have studied for A-level examinations in order to gain entry to university. However, this is simply not happening. In particular, it is not happening for the physical sciences of physics and chemistry. For example, Smithers and Robinson (2009) report that between 1982 and 2006, A-level physics entries halved from 55,728 to 27,466, albeit with some small recovery since 2006 with an increase of 7.3 per cent to 29,436. Beyond the United Kingdom, a number of countries do not appear to be doing much better. Again, Smithers and Robinson (2009) report that, at the A-level stage, take up has been falling in Australia, Eire, Finland and New Zealand. In European countries, concern has been expressed in a Nuffield report that 'many countries are experiencing significant problems with engaging students with the advanced study of physical sciences' (Osborne and Dillon 2008: 13). However, the report notes that this pattern is not universal across Europe, but appears to be correlated with the level of economic advancement of countries.

This situation of lower numbers of students studying physical sciences at an advanced level might trigger questions about why. There is strong evidence from a number of research findings that attitudes towards science in schools are negative. This needs to be clarified: it is not necessarily that the attitudes to science itself that are negative but that attitudes to *school science* are the problem. The Relevance of Science Education (ROSE) project survey of thirty countries found that attitudes to science and technology among adults and young people are mainly positive. However, answers to the survey show that although results vary strongly between countries, for European countries and Japan, the indications are that school science fails in many ways (Sjøberg and Schreiner 2010; see also ROSE n.d.), that science is 'important but not for me' (Jenkins and Nelson 2005: 41). Cultural factors may well be part of this downturn. The same Nuffield report (Osborne and Dillon 2008) indicates that the perceived values associated with science and technology are at odds with the values of contemporary youth and youth identity.

So how does school science fail? In examining this problem it may be useful to go back to the introductory conversation to highlight some of the less favourable attitudes shown by students. The example of the blast furnace almost provides a summary of the common features about students' negative attitudes to school science. These features, reported by Lyons (2006) and based on studies in Australia, Sweden and England, include views that the secondary school science curriculum is perceived as difficult, lacks personal relevance and utilises a pedagogy that is regarded as one that relies on the transmission of a body of facts. The lack of personal relevance is demonstrated in this transcript from the study by Osborne and Collins (2000: 22):

Roshni: The blast furnace, so when are you going to use a blast furnace? I mean, why do you need to know about it? You're not going to come across it ever. I mean look at the technology today, we've gone onto cloning, I mean it's a bit away off from the blast furnace now, so why do you need to know it?

Perceptions of difficulty and teaching style were shared by students in a Key Stage 3 (11–14 years) study conducted by Parkinson *et al.* (1998), where students referred to a dislike of copying, the amount of theory they needed to learn, and teacher factors that included too much teacher talk and unclear explanations. The Nuffield report (Osborne and Dillon 2008) summarises a lack of engagement with school science as a mix of: a lack of perceived relevance; a failure to generate a sense of anticipation; a pedagogy that lacks variety; less engaging teaching in science compared with other school subjects; content that is too male-oriented; and an assessment system that encourages rote and performance learning.

This situation appears to be a particular problem during secondary education. Research reported by Simon (2000) indicates that positive attitudes to school science peak at age 11 and then decline significantly. The view from the primary school is of secondary science being carried out in laboratories with sophisticated apparatus, of doing 'real' science (Jarman 1993). However, when they reach the secondary school, students have to come to terms with the reality of science that disappoints: a grey and dreary landscape.

What do students want?

With the problems discussed in the previous section, it is important to look at what the research says about the kinds of approaches students favour. Work on the 'student voice' is essentially about school students being consulted about teaching and learning, as this is an important prerequisite for informing teaching practices in schools (Smith *et al.* 2005). Rudduck and McIntyre (2007) suggest that four criteria seem to be used by students to explain their views about what makes for good lessons or helps classroom learning: the avoidance of tedium; the pursuit of meaningful learning; the need for togetherness; and the aspiration to be autonomous. Students responding to the Planet Science survey (Cerini *et al.* 2003) commented that they would like varied and interactive lessons, an opportunity to put forward their own ideas and the inclusion of controversial issues.

It appears that practical work features highly in students' attitudes towards, and enjoyment of, school science. Jarman (1993) reports that when students transfer from primary to secondary school they note that there is more practical work. Wellington (2005: 103) reports that when small groups of students were asked about their experiences of practical science they mentioned 'fun to do something, not just writing', 'all working together' and 'more interesting than reading or listening to teachers'. Osborne and Collins (2000) report reasons for students' interest in practical work as personal autonomy, fun, and science being more accessible and easily retained. Similar reasons were reported with the Planet Science survey (Cerini *et al.* 2003), where 47 per cent of students considered it made understanding easier, 30 per cent reported greater enjoyment and 12 per cent thought it provided deeper understanding. However, Abraham's work showed that only a small minority (8 per cent) of the students claimed that practical work helped them to learn, understand and recollect ideas and concepts, and students' preferences for practical work were more to do with situational interest and as an alternative to other methods of teaching. Similarly, Lindahl's research in Sweden (Lindahl 2003) reported that students feel it is never made clear why they need to learn content or do lab work, but they nevertheless want more variation in teaching.

What does the science curriculum want?

The 2010 Schools White Paper for England states that 'The curriculum should embody rigour and high standards and outline a core of knowledge in the traditional subject disciplines' (Department for Education 2010: 42). But what does this actually mean? What is 'rigour'? Is this the study of science content, the substantive concepts, laws and theories that are the backbone of traditional science? It could be argued that it is precisely these that are perceived as difficult, lack personal relevance and utilise a pedagogy that relies on the transmission of a body of facts, and so failing to engage students. Are 'traditional subject disciplines' those that in the past favoured a science curriculum focused on the elite minority, the very few who will go on to be the scientists and technologists of the future? Or should it provide a science for all? These are enduring debates.

Why is science for all important? Questions remain about why students need to make sense of the world and what science they need to do so. Despite problems with definitions, answers to these questions are often referred to as scientific literacy, science for citizens or the public understanding of science. A number of authors (e.g. Driver *et al.* 1996; Jenkins 1999) have discussed reasons for the importance of promoting a public understanding of science. The first of these is an economic argument which puts forward a direct connection between a public understanding of science and a nation's wealth. A second argument is the democratic one which claims that understanding science is a necessary part of an individual's participation in the decision-making processes of society. Jenkins (1999) notes that many countries are revising or reforming school science for an informed citizenry, although the impulse for doing so is often principally economic rather than democratic. Third, the utilitarian argument suggests the practical usefulness of science for living in a modern society where decisions are needed about health and welfare. A fourth argument concerns social considerations about maintaining cohesive links between science and a wider culture in order that science is not seen as a remote discipline alienated from the social world.

One essential part of scientific literacy is an understanding of the nature of science itself – how scientists arrive at knowledge. It is about the process of how scientific knowledge is built up, the means used to develop ideas by observing, thinking, experimenting and validating, and the wider social aspects of science. Although there is no fixed scientific method that scientists always follow, scientific enquiry demands evidence that may be obtained in different ways. It blends logical thought and imagination to explain and predict phenomena; it attempts to identify and remove sources of bias; and its findings are open to scrutiny, criticism and debate. The scientific enterprise itself is not one of the lone scientist at the bench, but a complex social activity that may be conducted across subject disciplines and in different institutions. Scientists follow accepted – and changing – ethical principles and procedures, and are able to participate as both specialists and citizens in the public domain.

We live in a society which is reliant on science and technology. It is therefore desirable, if not essential, for everyone in the society to have an understanding about the science and, importantly, an understanding of the impact that science has on society, in order to make informed judgements. School science contributes to public understanding by developing students' understanding of the scientific enterprise itself, its aims and purposes and the knowledge produced. This understanding is needed in order to develop an appreciation of the power and limitations of scientific claims, and to produce informed citizens who can participate in a modern democracy. As the 2003 PISA report notes:

> An important life skill for young people is the capacity to draw appropriate and guarded conclusions from evidence and information given to them, to criticize claims made by others on the basis of evidence put forward, and to distinguish opinion from evidence-based statements.
>
> (OECD 2007: 132)

This does not mean that the 'big ideas' of science are ignored: concepts and explanatory frameworks are important in order to understand science in society. Some knowledge of topics such as particles, forces, and the relationships between living things and their environments are still needed – clearly an understanding of climate change would not be easy without some understanding of combustion and photosynthesis. However, students' engagement with the issues of the world around them – a number of them controversial – is important. It may be useful to consider the extract in Box 1 below, taken from a news article about genetically modified (GM) crops.

Box 1: GM crops

In a BBC News report from The Environmental Audit Committee of MPs, chaired by Joan Walley, MP for Stoke-on-Trent North, the committee said an independent body should be established to research and report on the potential impact of GM crops on the environment, farming and global food system. 'Until there is clear public acceptance of GM and it is proven to be beneficial, the government should not license its commercial use in the UK nor promote its use overseas', Ms Walley said (BBC News 2012).

This example might be used to consider the evidence for and against GM crops, their economic and environmental impact, and their human benefits. It highlights some of the important components of scientific literacy, including the need to separate scientific evidence from personal views and anecdote, the ways in which scientists work in laboratories, in collecting data, in reaching conclusions and the tentative nature of conclusions. It may enhance students' ability to discuss, to weigh the evidence and to support or refute arguments. It may encourage students to engage with some of the knowledge and understanding about the big ideas of science. It certainly does not lack rigour.

There are enduring debates about what science education in schools is for. Is it to educate a few per cent of the population for careers in science and technology? Or is it part of a general education for the majority of students who will not go on to study science at a more advanced level, but who need to be able to respond to the impact of scientific and technological changes as they throw up important social, economic, political and environmental issues?

Changes to the science curriculum originated at the instigation of prominent science educators with a series of open meetings in 1997 and 1998. These culminated in the document *Beyond 2000: Science education for the future* (Millar and Osborne 1998) that was a product of a desire to provide a vision of science education that addressed the needs and interests of young people as future citizens at the end of the twentieth century. One important outcome from *Beyond 2000* was the inception of the Twenty-First Century Science project that was the first attempt to develop and pilot a major curriculum initiative and to use evaluations of the pilot to inform further development. This new curriculum attempted to overcome some of the criticisms of earlier curricula by being less prescriptive and including relevant and a little contemporary science. The lessons learned from the introduction of this new curriculum were, in turn, to influence the 2004 Science National Curriculum for England (DfES/WO 2004) by slimming down the content and including *How Science Works*. The four main sections of the *How Science Works* strand of the National Curriculum are summarised in Box 2.

Box 2: How science works

- Data, evidence, theories and explanations including: data collection, analysis and interpretation; development of theories; explanations using scientific theories, models and ideas, and questions that science cannot answer.
- Practical and enquiry skills involving planning, testing ideas, data collection, safe working and evaluation.
- Communication skills using qualitative and quantitative approaches, presenting information, developing an argument, concluding and using scientific, technical and mathematical language, conventions and symbols and ICT tools.

(Continued)

> *(Continued)*
>
> - Applications and implications of science that deal with: benefits, drawbacks and decisions; ethical issues; social, economic and environmental effects of decisions; and how uncertainties in scientific knowledge and ideas change over time.

Implementing *How Science Works*

This section discusses findings from a recent longitudinal research project that sought to explore pre-service or trainee teachers' views about how *How Science Works*, as a major curriculum change in schools, was being implemented in their school experience secondary schools (Toplis *et al.* 2010). We posed two main research questions: Is there a paradigm shift in terms of a change in the pattern of science education, with a move from teaching facts to understanding concepts? (This would represent a change in thinking, or 'mind set'.) Is there an associated change in pedagogy in terms of (1) mind set, (2) methods and (3) materials? The study took place with three (later two) universities in the London area over a three year period with 70, and later 50, trainees over the study period. All the trainees studied a one-year course that included university-taught elements and two blocks of observation and teaching experience in schools. We regarded these trainee teachers to be in a unique position to collect data during their school experience, as they had become participants in their communities of practice in schools. As participants, they had access within their schools to the students they teach and the work of teachers. They were able to benefit from the work of established teachers through observing lessons, meetings and feedback on their own supervised teaching in science classrooms. As the trainees came with a very limited experience of the history of school science under the National Curriculum, they could bring fresh insights without prior experiences or preconceived ideas about the science curriculum.

The study was carried out in two stages. Stage 1 involved five emailed questions four weeks before they completed their final period of school experience. The questions were related to the main research questions, with specific questions related to the three aspects of pedagogy in the second research question, namely:

Question 1   What do you understand by *How Science Works*? [Mind Set]
Question 2   Describe the resources that are being used at Key Stage 4 (age 14–16 years) in your present practice school. [Materials]
Question 3   What are your experiences of teaching approaches to *How Science Works* at Key Stage 4? [Methods]
Question 4   What challenges have you experienced with teaching *How Science Works*? [Methods]
Question 5   How do you see that pupils receive *How Science Works*? [Mind Set]

Stage 2 involved small discussion groups (three or four trainees) in the final weeks of their university courses. The discussion started with one of the five questions, where the trainees summarised their discussions, adding to the responses from each group, and repeating the process with all the questions. All emailed and discussion summaries were analysed collaboratively by the researchers to categorise emergent themes. Our findings are grouped below according to the emergent categories.

Trainees' understanding of How Science Works

This concerned investigation skills and the procedural knowledge needed to carry out practical enquiry tasks. Within this area were: planning, predicting and hypothesising, including sample size and an understanding of variables and their control; collecting, analysing and interpreting data; and evaluation, including reliability and validity. These components included most of the requirements for practical enquiry since the inception of the National Curriculum in 1989. These ideas and the terminology used have become familiar to science teachers and have now become part of the mind set of trainee teachers as they become members of their respective communities of practice. We did notice a trend towards the historical aspects of *How Science Works*, including famous scientists, discoveries and development of theories, as well as a growing understanding of the importance of communication that ranged from the use of scientific vocabulary and presentation skills by students, to putting everyday concepts into perspective and debates about how science affects society. The responses in this category indicate that trainees were being exposed to some of the underlying pedagogies of *How Science Works* that seek to question scientific ideas and present an argument, such as those underpinned by the research and development of argumentation (e.g. Simon *et al.* 2006).

There was an emerging recognition of the need for higher order thinking skills – skills such as analysis, synthesis, evaluation and creativity that appear higher in the revised Blooms Taxonomy (Anderson and Krathwohl 2001) – that relate to understanding science concepts rather than a recall of scientific knowledge.

Resources used

Over the time period, data showed that schools place a heavy reliance on published schemes that included textbooks (often specifically linked to examination board specifications); published worksheets and questions, and ICT materials that included multi-media CD packages such as simulations, web-based resources that included video clips and materials and presentations based on PowerPoint. There was very little evidence of teachers and science departments developing their own 'bespoke' materials.

Experiences of teaching approaches to How Science Works

Data indicated a strong dichotomy in response. In the few schools who had taken a determined approach, the changes had been positive. These included: an improvement in the way practical investigations were being presented and managed by teachers, with increased student involvement; an increase in the use of role-play, the use of animations and topic research to support conceptual understanding; and a new approach to develop imagination, participation and provide opportunities for creativity. With many schools, however, there were still areas where there was little evidence of progress. There was a continuing lack of help and support for teachers who were still unsure of what was required, a lack of resources, and an approach that continued to challenge lower ability students, whose understanding of vocabulary, lack of prior knowledge and the need for higher order thinking skills resulted in reduced interest and motivation.

Challenges experienced with teaching How Science Works

Some of the challenges involved with the practicalities of teaching *How Science Works* reduced over the three year period. These included the need to convince students that *How Science Works* was still science and making them aware of the aims and the links to science, for example, 'students used to being fed facts, many just want to sit and answer questions out of a textbook'. There were further problems with providing the right level of guidance for practicals and making sure all students were involved in discussions. This was seen as an inevitable consequence of the process of embedding curriculum change. However, there was a need for content to be discussed; the proportion of this category increased over the period. There was no change in the assessment-driven nature of the curriculum over the two cohorts. Student ability still remained a key factor and a challenge.

How students receive How Science Works

There was an increase in reports of engagement and enjoyment of students with science over the study period, especially so for the more able. Problems still remained in terms of differentiation. Evidence indicates that where higher ability students were challenged, they enjoyed science and responded well, particularly where higher order thinking skills were involved. However, where science was insufficiently challenging, their response was negative. Lower ability students found some of the work too challenging but responded well when it was tailored to their needs.

In conclusion, there is every indication that a new approach to teaching science through *How Science Works* is a positive step that engages, enthuses and challenges students, giving them a deeper understanding, interest and insight into the concepts and nature of science. Our findings indicated that how *How Science Works* is adopted in schools lies on a spectrum between what we refer to as creative and restricted implementation.

Schools that have engaged creatively in this process of transition have recognised the importance of developing historical context, communication and research skills, as well as the basic practical skills of experimental work. Many schools have not. They have continued to be hampered by a restricted view towards curriculum change characterised by a lack of student motivation, formal support and resources. This restricted implementation is understandable, given the pressures on the teaching profession that include financial constraints, assessment linked to success in national league tables, reduced in-service training, innovation overload and inertia. In order to benefit fully in this process, students need to be challenged into developing higher order thinking skills. Lower ability students can be left demotivated, with difficulty in developing vocabulary, debating skills, conceptual understanding and the ability to work independently. The underlying issue throughout is differentiation.

Conclusion: where do we go from here?

In order to bring this chapter together, it is appropriate to revisit Lyon's (2006) three points that highlight the main negative aspects of science education – relevance, difficulty and a pedagogy of transmission – and combine these with what students want. It may be worth noting that exactly what students want may not always be desirable within the constraints of schools as organisations or the aims of a curriculum, but they do provide a reasonable indicator about interest and engagement.

As our work on implementing *How Science Works* in the curriculum has indicated, creative and yet differentiated approaches to school science can both engage and challenge students. The question is about which pedagogies can encourage engagement and challenge and address the more negative aspects of existing science education.

School students are growing up in a fast-changing and technological world, and that world throws up a number of social, ethical and environmental issues. If school science is to be relevant to them, it surely needs to provide them with the tools to form opinions, use evidence and critically review reports. It also needs to be both contemporary and able to consider the sometimes controversial issues that arise. Importantly, school science has to use fundamental scientific approaches about the use of data, evidence and how data is obtained in order to provide reasoned arguments for addressing socio-scientific issues. Research and development in teaching and learning with different pedagogies has provided knowledge foundations about how these can be used in the classroom, and how they are received by both students and teachers. Levinson's work on teaching controversial issues has discussed the nature of these issues and highlighted the differences between the features of socio-scientific issues and features of school science, as well as providing contemporary and pertinent examples (Levinson 2011). Work on using evidence for argumentation and how argumentation can be taught and managed in the classroom has received attention in the United Kingdom and other European countries (Osborne *et al*. 2004). The late Phil Scott's legacy of work on dialogic teaching has focused on the talk of science classrooms and in the different kinds of interactions between teachers and students that contribute to meaning-making and learning (Mortimer and Scott 2003). Work on context-based approaches to teaching A-level sciences, often perceived as difficult, have brought science up to date and gained popularity with both students and teachers (e.g. Bennett and Lubben 2008). Scientific inquiry (or enquiry in the United Kingdom) has received a good deal of attention in the United States and more recently in Europe (e.g. Rocard Report 2007) as an approach that can lead to greater engagement of students through practical work, model some of the ways that scientists work and provide students with a sense of autonomy and involvement. One question could be: Why is inquiry not as widespread as it could be?

Invariably, findings from research, development and evaluation on science education pedagogies will take time to bed down in practice. If their adoption is to become widespread, it will rely on the creativity, imagination and, importantly, dissemination by and to science teachers and science educators if a grey and dreary landscape is to brighten up.

Key questions

I have taken some of the questions already discussed in the chapter, and invite you to answer these for yourself:

1   What do you understand by the phrase 'how science works?' Imagine you have to explain the workings of science to your (non-scientific) mother.
2   What resources do you use at Key Stage 4 in your present practice school? Any mentioned here?
3   What are your experiences of teaching approaches to *How Science Works* at Key Stage 4? Reflect a little on what you have done (if anything!) and how it went.

Further reading

Millar, R., and Osborne, J. (1998) *Beyond 2000*. London: King's College London, School of Education. This report discusses the kind of science education we may need for the future and how it might be achieved. It provides ten recommendations that relate to both scientific literacy for all and science education for those who wish to pursue science further. It can be downloaded from the Nuffield Foundation website: http://www.nuffieldfoundation.org/sites/default/files/Beyond%202000.pdf

Toplis, R. (ed.) (2011) *How Science Works: Exploring effective pedagogy and practice*. London: Routledge. This book provides a collection of chapters that include: the school curriculum; students' views; how scientists work; teaching controversial issues; questioning in science; investigative science; argumentation; the role of ICT; and teaching outside the classroom.

References

Abrahams, I. (2009). Does practical work really motivate? A study of the affective value of practical work in secondary school science. *International Journal of Science Education*, 31(17): 2335–2353.

Anderson, L. W., and Krathwohl, D. R. (eds) (2001) *A Taxonomy for Learning, Teaching and Assessing: A revision of Bloom's Taxonomy of educational objectives: Complete edition*. New York: Longman.

BBC News (2012) MPs call for wider food strategy (13 May). Available online at: http://www.bbc.co.uk/news/uk-18046179 (accessed 1 August 2012).

Bennett, J., and F. Lubben (2008) Context-based chemistry: The Salters approach. *International Journal of Science Education*, 28(9): 999–1015.

Cerini, B., Murray, I., and Reiss, M. (2003) *Student Review of the Science Curriculum*. London: Planet Science, Institute of Education and the Science Museum.

Department for Education (2010) *The Importance of Teaching: The Schools White Paper 2010*. London: The Stationery Office.

DfES/WO (Department for Education and Science/Welsh Office) (2004) *Science. The National Curriculum for England*. London: HMSO.

Driver, R., Leach, J., Millar, R., and Scott, P. (1996) *Young People's Images of Science*. Buckingham: Open University Press.

Jarman, R. (1993) 'Real experiments with Bunsen burners': Pupils' perceptions of the similarities and differences between primary science and secondary science. *School Science Review*, 74(268): 19–29.

Jenkins, E. (1999) School science, citizenship and the public understanding of science. *International Journal of Science Education*, 21: 703–10.

Jenkins, E. W., and Nelson, N. W. (2005) 'Important but not for me': Students' attitudes towards secondary school science in England. *Research in Science and Technological Education*, 23(1): 41–57.

Levinson, R. (2011) Teaching controversial issues in science. In R. Toplis (ed.) *How Science Works: Exploring effective pedagogy and practice*. London: Routledge (pp. 56–70).

Lindahl, B. (2003) Pupils' responses to school science and technology? A longitudinal study of pathways to upper secondary school. *Acta Universitatis Gothoburgensis*. Available online at: http://gupea.ub.gu.se/dspace/bitstream/2077/9599/1/Summary.pdf (accessed 26 December 2009).

Lyons, T. (2006) Different countries, same science classes: Students' experiences of school science in their own words. *International Journal of Science Education*, 28(6): 591–613.

Millar, R., and Osborne, J. (1998) *Beyond 2000*. London: King's College London School of Education.

Mortimer, E. F., and Scott, P. H. (2003) *Meaning Making in Secondary Science Classrooms*. Milton Keynes: Open University Press.

OECD (2007) *PISA 2006: Science Competencies for Tomorrow's World. Executive summary*. Available online at: http://www.pisa.oecd.org/dataoecd/15/13/39725224.pdf (accessed 23 August 2009).

Osborne, J., and Collins, S. (2000) *Pupils' and Parents' Views of the School Science Curriculum*. London: Kings College London.

Osborne, J., and Dillon, J. (2008) *Science Education in Europe: Critical reflections.* London: The Nuffield Foundation.

Osborne, J., Erduran, S., and Simon, S. (2004) *Ideas, Evidence and Argument in Science.* In-service Training Pack, Resource Pack and Video. London: Nuffield Foundation.

Parkinson, J., Hendley, D., Tanner, H., and Stables, A. (1998) Pupils' attitudes to science in Key Stage 3 of the National Curriculum: A study of pupils in South Wales. *Research in Science and Technological Education*, 16(2): 165–76.

Rocard Report (2007) *Science Education Now: A renewed pedagogy for the future of Europe.* Luxembourg: Office for Official Publications of the European Communities.

ROSE (Relevance of Science Education) (n.d.) Website at: http://www.uv.uio.no/ils/english/research/projects/rose/ (accessed 25 June 2013).

Rudduck, J., and McIntyre, D. (2007) *Improving Learning Through Consulting Pupils.* Abingdon, Oxon: Routledge.

Simon, S. (2000) Students' attitudes towards science. In M. Monk and J. Osborne (eds) *Good Practice in Science Teaching: What research has to say.* Buckingham: Open University Press (pp. 104–19).

Simon, S., Erduran, S., and Osborne, J. (2006) Learning to teach argumentation: Research and development in the science classroom. *International Journal of Science Education*, 28(2–3): 235–60.

Sjøberg, S., and Schreiner, C. (2010) *The ROSE Project: An overview and key findings.* Available online at: http://roseproject.no/network/countries/norway/eng/nor-Sjoberg-Schreiner-overview-2010.pdf (accessed 4 August 2012).

Smith, C., Dakers, J., Dow, W., Head, G., Sutherland, M., and Irwin, R. (2005) A systematic review of what pupils, aged 11–16, believe impacts on their motivation to learn in the classroom. In *Research Evidence in Education Library.* London: EPPI-Centre, Social Science Research Unit, Institute of Education, University of London.

Smithers, A., and Robinson, P. (2009) *Physics Participation and Policies: Lessons from abroad.* Buckingham: Carmichael Press.

Toplis, R., Golabek, C., and Cleaves, A. (2010) Implementing a new science National Curriculum for England: How trainee teachers see the *How Science Works* strand in schools. *The Curriculum Journal*, 21(1): 65–76.

Wellington, J. (2005) Practical work and the affective domain: What do we know, what should we ask, and what is worth exploring further? In S. Alsop (ed.) *Beyond Cartesian Dualism.* Dordrecht: Springer (pp. 99–109).

Part III

Classroom debates

Chapter 7

Online gaming and digital fantasy for scientific literacy

Nic Crowe

Introduction: (no) games in the classroom

It is a cold, wet Wednesday afternoon and I am sitting in a draughty porta-cabin classroom surrounded by robots. Not those yellow car-constructing mechanicals from the FIAT adverts, or the Terminator-style, laser-packing, planet-invading synthetics so beloved by Hollywood, but robot models – or more correctly, robot prototypes. There is a large humanoid-looking hulk that looks suspiciously like a Gundam (a piloted mechanical figure popular in Japanese animation) all glowing eyes and military-influenced paint, a small squat 'droid' – reminding me of a stunted Dalek – that bleeps and whirs, and a small spider-like creature with nine legs that scuttles around the floor end-lessly bumping into chair legs. What is interesting is not the robots themselves, but the children controlling them. These low-tech surroundings house the 'Science Club', a group of about twelve year 9–11 students that meet weekly after school. Nothing particularly strange about that, except that this is a project aimed at engaging pupils who are disenfranchised from mainstream school. Normally it is a struggle to get such pupils to attend, let alone engage in extra-curricula activity. Scott, a year 10, excitably explains the club's appeal:

> Look at all this, where do you get to do crazy stuff like this … I play this [computer] game, 'Portal'. In it you got to solve problems by opening and closing portals that you shoot on the wall with your portal gun. Yeah, I know it sounds simple, but not every wall opens up so you have to think ahead, work out the correct angle to shoot at, and to swing or jump from, or the speed you fall at, to time a shot. Some things only work if you do them in the right order so you have to think about this as well. Mr Dickens plays too, so we learnt in his class that science problems were the same. These robots ain't as cool as the 'Aperture Science bots', but like Chell [the player-character in the game] you gotta think about how you manoeuvre them around the room. How they walk and stuff. It's just another puzzle really. Me and Vil come here cause it's just like the game. You get to play and you forget that you are learning. That's right innit Vil?

Vil looks up from the robot game she is playing.

> Yeah, there ain't no Mecha in maths class that's for sure. [A Mecha is a slang term for robots, derived from the large mechanicals popular in Japanese Anime and Manga culture.]

Scott taps her on the head:

> 'How would you know, you never go!'

Vil re-arranges her hair, grins at me and quips:

> And now you know why!

(Field diary extract, 2010)

In 1958, William A. Higginbottom, an engineer who had worked on timing devices for the atomic bomb, produced a simple oscilloscope-based version of 'tennis' as a means of entertaining visitors to the US Nuclear Research Facility where he worked. Three years later, MIT computer scientist Steve Russell created the first true digital game 'Spacewar', as a side project to showpiece the technical capabilities of his DEC PDP-1 Mainframe. So, digital games are historically linked to the activities and thoughts of the scientific community (Squire and Patterson 2010). Perhaps we should not be too surprised, then, that Scott and Vil have discovered an interest in science through their computer gaming. In the early days of their development, these games continued to be little more than a technical 'cult' within a tight scientific community (Poole 2000), though they have since cast aside these shackles and become a central aspect of contemporary popular culture (Oblinger 2004; Crowe and Bradford 2006). Computer games are popular. Young people have always been early adopters of this type of technology (Raine and Horrigan 2005) and digital games are an important aspect of young people's leisure (Livingstone and Bober 2005). Regular game usage in the United Kingdom sits at 91 per cent in the 12–15 year age bracket, rising to 94 per cent amongst 8–11 year-olds (OFCOM 2011), and sales of digital titles now exceed those of both books and popular music (Pricewaterhouse Coopers 2010). Faced with such statistics, the follow-up to the Government-commissioned Byron Review (Byron 2010) had little choice but to recognise that engaging with digital technology – including gaming – was becoming an ordinary aspect of childhood. Yet, against this backdrop of domesticity, educators in the United Kingdom have remained reluctant to use digital games in the classroom.

The educational role of popular technologies has always generated fierce debate. Supporters of digital platforms (e.g. Prensky 2001; Gee 2007) denounce the teaching profession for hesitating to embrace new or unfamiliar technology, while critics mistrust the introduction of popular media into the classroom fearing that it imposes an 'entertainment modality' on learning (see Postman 1985). The games industry has responded by stressing the 'edutainment' and educational strands to their developments and, whilst this initially found support among 'progressive educational reformers pursuing equity in learning' (Ito 2009: iv), they have since been accused of 'failing to integrate content and game play, having poor production values, and generally dumbing down for educational audiences' (Squire and Patterson 2010: 4). Yet Friere (1985) reminds us that lived contemporary experiences are an important aspect of learning, and Gee (2003) wryly concludes that, in contrast to the ineffectiveness of much technology-based approaches, digital games continue to develop a reputation for being fun, engaging and immersive, requiring deep thinking and complex problem solving. It is perhaps easy, then, to dismiss digital games as lazy forms of popular entertainment, but to do so would be to miss out on novel modes of pupil engagement (Crowe and Flynn 2013).

In this chapter I want to explore these debates, and reflect on what opportunities digital games might offer to the teaching of science in schools. There is an emergent body of work (particularly in the United States) that has begun to scrutinise the educational possibilities proposed by computer games. Yet there has been less written on how pupils feel about their learning in such scenarios. Interestingly, Morrow (2005) observes that, in too many cases, research is something that is done *to* young people rather than *with* them. With this in mind, I will be drawing on material from an ethnographic study of digital play to illustrate how students experience their science learning and, more broadly, asking whether computer games can form the basis of effective and purposeful science education-based interventions.

Science (il)literacy

Science education is often presented as a central element of informed and engaged citizenship (Rutherford and Ahlgren 1989). While in the United Kingdom this notion of citizenship remains something of an abstraction, the concept is usually actualised in the idea of scientific literacy (Laugksch 2000). The argument put forward is simple. Society has become reliant on science and technology, and is now arguably understood best through a scientific–technical discourse. It is therefore essential for everyone to have an understanding, not only about the science/technology itself, but more importantly its impact on our everyday world (Jenkins 1999). Only through such understanding can we make informed judgments about the power and limitations of scientific knowledge and become informed citizens who can fully participate in a modern, technical and democratic society (Roth and Désautels 2004). This utilitarian approach challenges the long-held belief that science is a subject for specialists with no importance at all for the ordinary citizen (Roth and Désautels 2004), but perhaps more significantly it also presents science as a 'problem-solving multidisciplinary activity done by professionals' (Leite *et al.* 2007: 1).

Although there is a continuing tension between a need to study topics that are relevant to pupil's everyday lives and the more formal programmes that focus on key concepts and knowledge (Toplis 2011), there is a growing rejection of studying science 'for science's sake'. Of course, it has long been recognised that education is more than the mere transfer of information or knowledge (Mezirow 2000). However, scientific literacy, as described here, requires an educational approach that addresses critical thinking skills, particularly problem-solving and those associated with independent learning. The 1999 PISA 'New Framework for Assessment' begins its section on 'Scientific Literacy' with the following observation:

> An important life skill for young people is the capacity to draw appropriate and guarded conclusions from evidence and information given to them, to criticise claims made by others on the basis of the evidence put forward, and to distinguish opinion from evidence-based statements. Science has a particular part to play here since it is concerned with rationality in testing ideas and theories against evidence from the world around.
>
> (OEDC 1999: 59)

Science education, along with technology, is seen as being central in preparing all students for lifelong learning (Fullick 2004). It requires the teaching of appropriate lifelong learning competences, so that when young people leave school they can continue to update their science knowledge by themselves and also use new knowledge in their individual and professional lives (King 2001). Arguably, this approach is most effective when educators are able to make clear links with the contemporary experiences of young people, and use this as the starting point for intervention (Davies 2005).

The UK Government certainly thought so: 'while not everyone is in the business of science, science is everybody's business' (Brown 2009). This became a central theme of New Labour initiatives to enliven an increasingly stagnant pool of science, technology, engineering and mathematics (STEM) professionals, by increasing the supply of young people into the profession. However, even with this utilitarian emphasis on grounding science in the everyday lives of its participants, as other authors in this book suggest, science education still does not capture the imagination of the students themselves. Despite being a compulsory subject until Year 11, science subjects have failed to be a popular choice at post-compulsory levels. And, as Joanne Cole suggests in Chapter 4, since the 1980s, the percentage of students pursing science or science and mathematics post-16 has declined by more than half (Butt *et al.* 2010; Smith 2011). Behind these statistics is a growing unhappiness with science, as a group of young people from Surrey explained to me in 2011:

> Dull! Dull! Dull! And you *still* have to do it!
>
> (Vikash, Year 10)

Who is going to make it interesting, Nic? We have to do it anyway. So who is going to bother? I used to watch those adverts on TV for teachers and stuff. There was this one where the science teacher took them outside and made this wind tunnel with them. And I thought, not in my f****** school, science is shit there. We could do stuff like that but instead we sit in the lab full of equipment we never use, and copy stuff out of books. You tell me how is that exciting?

(Martin, Year 10)

Martin's point is well (and forcibly) made. The potential to make science attractive is often not realised due to poor teaching strategies rather than inadequate resources (Smith 2005). Martin's concern about being trapped in this system echoes similar observations made in the late 1960s:

If science were compulsory it must be attractive, if it is not attractive it will only suffer if made compulsory; and if it were attractive, it would not need to be compulsory.

(McPherson 1969)

Year 11 student Kerry describes the vicious circle outlined by McPherson in a slightly different way:

We don't like science, we only do it because we have to. Some of the kids mess about because they don't want be there. So Sir makes us copy stuff out in silence which pisses us off even more, so we mess around even more. We never get to do anything fun because he says he can't trust us, but can't he see that we only do it because he makes it boring in the first place! It's a shame because there are lots of interesting things that we could learn if only he would make it fun.

Kerry's final point is interesting. There is evidence from recent research undertaken by the Wellcome Trust (2011) that pupils find science an exciting and enjoyable area of study when it is done well. She also reminds us of the importance of teacher enthusiasm in determining engagement. Butt *et al.* (2010) observed that having a 'good' teacher encouraged pupils to learn science, whilst 'bad' teaching had a more noticeable negative effect than in other subjects. Wellcome concluded that:

Teachers who acted as a positive influence on attitudes towards science were reported to be those who: made science lessons enjoyable, interesting and understandable through their passion for their subject ... Conversely teachers who acted as a negative influence on attitudes were those who put students down; did not offer enough help or encouragement; expected students to just copy down notes; or had problems with discipline.

(Wellcome Trust 2011)

As might be expected, engagement with science comes down less to *what* is being taught but rather *how* it is actually delivered (DeWit and Osborne 2008). The science curriculum has been criticised for being content heavy (Osborne and Collins 2001), and over-reliant on repetitive written work (Owen *et al.* 2008). As the earlier comments serve to illustrate, these are factors that many young people seem to find off-putting. It is timely then to consider new, perhaps radical, ways to engage students.

(Will you) prepare to qualify

Might digital games offer an effective and attractive mechanism to establish scientific literacy? Putting their popularity aside, there are compelling education-based arguments that suggest they might provide a viable pedagogic tool. Problem-solving is regarded as a key skill for lifelong learning (Hoskins and Frediksson 2008) and, as I discussed earlier, is a central theme of a scientific literacy framework. Similarly, pupils express a preference for more practical, hands-on activities, which they believe make learning science more interesting and subsequently easier to understand (Wellcome Trust 2011: 6). Exponents of using games in the classroom argue that digital play can teach users to become effective problem solvers (Crowe and Flynn 2013). In well-designed games, the player can only advance to a higher level (or unlock more desirable equipment or resources) by testing out a range of different approaches and strategies. So, for example, *Portal* (the 2007 title played by Scott and Vil in the introduction) requires players to negotiate a series of puzzles. These are solved by creating inter-spatial portals between two flat planes using a 'portal gun' and then tele-porting the central character, Chell, through the gateways. The game's modified physics engine allows momentum to be retained through these portals, which must be used creatively to overcome *Portal*'s challenges. This basic gameplay is enhanced by additional features: tractor beams, laser redirection, bridges made of light, and paint-like gels that give surfaces special properties, such as acceleration and the ability to manipulate gravity. *Portal* requires the player to ask questions such as: What happens if I do this? Where does this go? What do I need to do next? Successful completion of the game requires lateral thinking and the ability to solve increasingly complex problems. Since the game is developmental – it gets harder the further it goes – players are required to learn the skills and knowl-edge they require to solve the next set of challenges. One of the key aspects of the most complex games is that players need to learn to cope with the rigours of the virtual world, thus one of the functions of digital games is that users are being 'taught' to 'learn'. Vil and Scott explain further:

Vil: The game might start simple, so you learn the basics, but as you get into it, things start to get much harder. More complicated puzzles and stuff. You have to remember what you learn from the room before and then try to apply this to the new challenges. Sometimes it is just trial and error, but after I have got through it seems so obvious and I always think: why didn't I just do that the first time? I suppose I am getting better and I just don't realise it.

Scott: Yeah, Vil is right. You don't know that you are learning new stuff. As you get further in, you have to think about acceleration, gravity and angles or you just won't make it through … Mr Dickens taught us about this in science sessions, but it wasn't until we played this and I saw what it meant to the way I controlled Chell that I started to get it. It made it all come alive. Kind of real?

These young people touch on arguably one of the most important aspects of game-based learning. It offers the opportunity to experience situated simulations that might not be possible – or as accessible – using other more passive texts or forms. To a certain extent, it is this interactivity that give such digital arenas their added importance, not only because it facilitates informal access to information and knowledge, but also because it is the main cause for its motivational power that makes students feel willing to learn (Leite *et al.* 2007; Fullick 2004). *Portal* is far more than a series of abstracted puzzles, however. The game 'works' because Scott and Vil become heavily invested in its narrative, principally through the character of Chell, whose escapes they are attempting to facilitate. For something to have an impact at the level of fantasy does not mean that it has no educational significance (Crowe and Watts 2012). As Gee notes, good digital games teach pupils 'to solve problems and reflect on the intricacies of the design of imagined worlds and the design of both real and imagined social relationships and identities in the modern world' (Gee 2003: 48).

Learning with(out) games

Crawford acknowledges that the fundamental motivation for playing games is a desire to learn:

> Games are thus the most ancient and time-honored vehicle for education. They are the original educational technology, the natural one … In light of this, the question, 'Can games have educational value?' becomes absurd. It is not games but schools that are the newfangled notion, the untested fad, the violator of tradition. Game-playing is a vital educational function for any creature capable of learning
>
> (Crawford 1982: 16)

Play is an important developmental process through which children make sense of their contemporary experiences. It represents 'the imaginary, illusory realisation of unrealisable desires' (Vygotsky 1933), so, in the same way that *Portal* allows users to manipulate aspects of the physical environment, play offers opportunities to explore some aspects our world that might be normally impossible in everyday life. It is through such fantasy that children in particular come to develop a greater sense of meaning and purpose about their own lives (Bettelheim 1976). When educators structure learning in play, they are facilitating an engagement with fantasy activity that would not otherwise be possible. Digital play is therefore not an adjunct to learning but an integral part of this developmental process.

This type of fantasy-based learning is not simply inherent in children, however – it is culturally acquired and culturally developed through informed action. We think and understand best when we can imagine a situation and that prepares us for action. The simulated environments and processes of computer games provide the opportunity to think, understand, prepare and execute actions (Gee 2003). Digital arenas offer advantages over other forms of play in this respect. Computer games provide opportunities for continued practice because negative consequences are not typically associated with failure (Groff *et al.* 2010; Ke 2009). In a game such as *Portal*, failure serves as an integral part of the learning experience – indeed trial and error is an appropriate learning strategy across most computer games. Players only improve through repeated practice, as Rosie explains:

> You have to keep going if you want to get good. When you do Quests, there is really only two ways to get through it. Go to a walkthrough site, or just try things out until you get it right. Only a noob, uses a walkthrough. No one respects anyone who does that. The players who get the most respect are those that keep going, even if that takes a longer time [laughs]. No quick fixes on here.
>
> (Rosie, 14)

> Good games are challenging – I don't mean hard, I mean you have to think. You know, really think! Not just about how to solve immediate problems like how to get to the next level, but you also have to think long term … You have like a 'Skill Tree' that teaches you what to do, but only when you need it and when you are good enough at what you already know. So you learn to move, then run, then crouch, then jump, then double jump and different sorts of firing – like jump and fire … Then onto advance skills like co-op [with another player], crafting, trading. You don't learn stuff until you are ready. It's logical.
>
> (Vikkii, 14)

Rosie and Vikkii illustrate how, in digital play, children and young people create developmental environments that support them to do what might seem beyond them. They are encouraged to try new things alongside doing things that are familiar to them. The fantasy narrative requires them to push themselves ever forward, expanding their knowledge base and learning as a result. The girls also demonstrate how playing digital games can be a complex process involving a range of both short- and long-term strategies. Games need to be *learnt* and, for those prepared to persevere with in-game tasks, this is a considered and deliberate process in which short-term individual and emotional pleasures often need to be put aside for long-term success. Failure with limited consequence, agency and choice are seen as critical elements of a true gaming experience (Larson Mclarty *et al*. 2012).

Arguably one of the reasons that digital games have become so successful in challenging 'failure' in this way is that, like the best educational experiences, they are built on logical learning progression. As Vikkii showed earlier, games require players to learn in an ordered sequence, often directly linked to the immediate challenge that is facing them. Simple skills form the foundation on which more complex knowledge is subsequently based. Yet mastery of skills and knowledge are integral to advancing in the narrative and are often only acquired through repetition – by replaying the challenge:

> The only way to get through that bit with the crushing jaws is to jump and then fire your PG [Portal Gun] at just the right point. It takes practice – took me about 10 goes to get it just right – but when you have done it you feel you have earned your way into the next level. Job well done as my dad says. Games are like that, if you don't learn what to do, and learn it properly, then you will not advance.
>
> (Alfred, Year 11)

Alfred's sense of accomplishment at a 'job well done' is clear. There are echoes here of Rosie's earlier point about earning respect. Amongst these young people there is an expectation that the most accomplished players will have invested time in their learning rather than trying to take the easier option. A state of pleasant frustration – challenging but doable – is an ideal state for learning several content areas such as science (diSessa 2000). This learning process is further supported by the way that digital games weave scaffolding into their narratives – mainly through the mechanism of levels. Successful games 'set the level of difficulty at the point where the learner needs to stretch a bit and can accomplish the task with moderate support' (Jalongo 2007: 401), which then provides an effective match between learning and student need. It affords a support structure early on in the game – less complicated puzzles, simpler challenges – giving players the time and space to practice and learn more complex skills.

(Un)supported learning

The science adventure *Crystal Island* provides a more developed example of how scaffolding can be used in a gaming environment. Created using the same game engine as *Portal*, the narrative presents a science mystery about an epidemic that has struck a team of researchers living on a remote tropical island. Players explore the island to find and examine objects, talk with characters, and use the laboratory equipment (and other resources) to solve the mystery. As the investigation progresses, the player is required to complete an in-game diagnostic worksheet in order to record results, pose questions and ultimately a diagnosis. The organisation of worksheet offers an opportunity to scaffold the problem-solving process alongside other inquiry supports, such as dialogue trees and in-game messages and cues from characters. The game requires a variety of inquiry techniques and there is ample opportunity to 'test out' a range of approaches to problem solving. Players can talk to sick characters about their symptoms; consult books and leaflets to try to match what they have been told with known diseases. They can even call upon in-game 'experts' to discuss in more depth aspects of different infections. Once they have formulated a hypothesis they can test this using the virtual research lab (to test whether in-game objects have become infected) and use the diagnostic worksheet to organise their findings to locate the likely source of the infection (this can be constantly checked by the camp nurse).

One of the most interesting aspects of *Crystal Island* in educational terms is that it allows students to remain in control of their own learning. Unlike the somewhat linear approach of *Portal*, the game offers a more open-ended inquiry-based narrative. Whilst both games allow students to learn independently, there is an increased sense of ownership and agency in this sort of open world. Since lack of agency is one of the most cited criticisms of E-Learning (Dalton 2000) it is significant that digital games encourage active learning and choice. Like *Crystal Island*, the historical science-based game *River City* offered an independent environment in which students conducted experiments and created hypotheses to solve science-based challenges (Ketelhut *et al.* 2006). This approach affords students the freedom to exercise a choice of learning styles and engage in a more active approach to their learning. This is understandably attractive, particularly given the 'closedness' of much of the curriculum (Klopfer *et al.* 2009):

> For once it is nice to be left to do things for yourself, how I wanted to do it you know. We weren't just dumped there and left to get on with it, there is plenty of help in-world – clues and directions and stuff – but the whole idea is that you do things in a way that you want to. It is part of the learning I guess.
>
> (Josh, Year 10)

> It was good to be left to try to figure things out for yourself. I really liked talking to the different characters about their symptoms. Some of them would tell you one thing and others would make how they were feeling sound worse than it was. You have to use the sheet to record things they say. The nurse at the hospital helps you to decide who is telling you the truth. … What did I learn? That different people describe the same symptoms in very different ways. You can't always trust the first thing you are told and you need to look further.
>
> (Peter, Year 9)

Peter makes an interesting point. One of the features of digital games is that they often present a multi-faceted approach to problems. Rather than simply resolving a challenge, the student is presented with a whole new level of inquiry, as the opening of multi-disciplinary game *Quest Atlantis* reveals:

> The Arch will not answer your questions, but leave you with better questions. Questions that you will spend a long time answering, so that later you can discover your own answers.

This game is explicit in its emphasis on independent learning. There are no teachers to provide immediate answers in *Quest Atlantis,* rather 'a network of peers and mentors' (Squire and Paterson 2010: 15) who facilitate a journey of self-discovery. The enigmatic introduction also serves to draw students into the narrative:

> I was like, what the f***? It sounded kinda mysterious so I thought, shit, yes I'll play! There is actually a lot of cool stuff here to help with various subjects, but unlike lots of these schooly games it was interesting from the start. It made me want to play it, not just because Miss said we had to.
>
> (Jack, 14)

Jack shows us the importance of narrative in establishing student interaction with the game. In the same way that Scott and Vil were drawn into the world of *Portal*, so games such as *Quest Atlantis* and *Crystal Island* use fantasy to create the environment for a more satisfying and engaging learning experience. Students appear to be far more engaged in a learning environment when it is based around a coherent story-based narrative (Barab *et al.* 2005). Fantasy not only provides a context but also a structure around which the various in-game tasks can be organised. This offers students a focus for their learning activities; in *Portal* they try to help Chell escape the research facility, in *Crystal Island* they must discover the sources of the infection before the inhabitants all die. The narrative is an important motivation tool in this respect, since young people are more motivated when they feel a sense of personal attachment to a goal (Crowe 2011).

Good digital games can be seen as narrative 'microworlds' that provide immersive arenas within which play – and hence learning – can take place (de Freitas 2006). Csikszentmihayli (1990) puts forward the concept of 'flow', which he identifies as the complete engagement in an activity. Many of the elements of flow are well supported in digital games; a balance between ability and challenge, clear goals and feedback, a sense of control, immersion in the activity. Enjoyment is a central theme of this approach. It was clear how *Portal* helped aspects of learning come 'alive' for Scott: it taught him that science could be an extension of his gaming interests. Gee and Hayes (2009) extend flow in their discussions of Affinity Spaces to identify arenas in which young people congregate around common passions. This positivity can then be directed to facilitate curriculum-based learning:

> I liked Resident Evil – you know, the zombie virus game. When we played Crystal Island I thought that it was a little like that ... But it was more interesting because instead of killing the infected, you had to work out how to cure them and where the infection had come from in the first place. I wish the 'Resi' games had a bit of that. It was really complicated and good fun when you got into it. I didn't think I would like all the investigation and stuff but it turned out to be the best part of the lesson. Crystal Islands is a cool game, I wrote about it in English. I got a 'B' – my first one [laughs].
>
> (Jordan, 13)

Enjoyment leads to engagement and student engagement is in turn strongly associated with student achievement (Shute *et al.* 2009). The enthusiasm with which Jordan approached *Resident Evil* provided him with a doorway to an enjoyment of science learning that he had not experienced before, and even provided an opportunity to address learning in another curriculum area. Digital game-based activities have the potential to offer greater levels of engagement than might be found with regular classroom tasks (Rieber 1996). Of course, educators have always employed a variety of methods to keep students interested in the classroom, but for the young people referred to throughout this chapter, it seems to be more than just this. Arguably, it was their affinity with the medium outside of school that sits at the heart of their engagement.

Conclusion: you can (re)boot

If scientific literacy is to remain the cornerstone of science-based education in schools, then it must focus not just on science concepts but also on the processes that lie behind scientific thinking and the use of them in everyday life (Roth and Désautels 2004). Digital games would seem to go some way to meeting this imperative, not least because, like scientific literacy, they are seen to address twenty-first century skills (Gee and Shaffer 2010). The domesticity of gaming technology affords it advantages in terms of both learning skills and education environment. Of course games are just one of many strategies that can be deployed, and popularity on its own is not a pre-cursor to educational attainment. However 'positive reaction may not ensure learning, but negative reaction almost certainly reduces the possibility of its occurring' (Kirkpatrick and Kirkpatrick 2006: 30). Some of the examples cited in this chapter show what can be achieved when educationalists take an interest in the contemporary experiences of young people, and then attempt to incorporate those technology-based experiences into science teaching and learning.

Fantasy and narrative have an important part to play in creating the environment, within which such experiences can flourish:

> The games are really just like books, but books that you can live in yourself ... which I think helped me understand things when we had to read in class. I would pretend the character was one in a game that I controlled. It helped me to think through why they might have done things like that, or why things happened the way they did. It's just more interesting than words on a page!
>
> (Emily, 13)

Arguably both *Portal* and *Crystal Islands* 'work' because the environment is experienced through the first person. The student is placed directly into the learning situation – the experts and teachers speak directly to them – and this facilitates a deeper learning experience than a moot third person (Dickey 2005). But as Emily shows us, the novelty of a more accessible form of narrative is also a factor in the ease with which these pupils transferred game-based skills to their studies.

Digital games have traditionally been caught in a tension between narratology (games as 'stories') and ludology (games as formal rule-based systems or 'simulations'). Most modern games allow players to modify levels of difficulty and other aspects of the game environment (such as characters and equipment), all of which help stimulate an insight into the ways that simulations are constructed. But it would be wrong to regard digital games as simple simulations of the 'real world'. Parker (2004) suggests that the 'realism' of games lies not just in the ability to simulate the material world as an aesthetic, but in its accurate re-creation of material social processes as well.

Digital games then have a role to play in the science classroom, perhaps not as an alternative to more traditional modes of teaching, but certainly as a pedagogic tool to complement them. The importance of digital games to the science curriculum lies in their ability to address a wide spectrum of learning styles within a complex decision-making framework (Squire 2006). Students are required to think systematically and take a relational approach rather than rely on isolated facts or ideas. The complexity of a gaming environment such as *Portal* or *Crystal Island* directs students not just to apply knowledge but more importantly adapt it to changing scenarios. This sits at the heart of scientific thinking, and is key to establishing scientific-literacy in the twenty-first century. Lev Vygotsky, believed that with new technologies come new human capacities and a need for new approaches to learning. 'The invention of new methods that are adequate to the new ways in which problems are posed requires far more than a simple modification of previously accepted methods' (Vygotsky 1978: 58). With this in mind, it is perhaps appropriate to end where I began – visiting Scott and Vil in the Science Club:

Vil pauses her game, puts down her controller and turns her attention back to her science textbook. She casually thumbs through the pages and then looks back at the monitor. A research facility robot is frozen in mid-air. She asks:

Do you know how to calculate velocity?

I shake my head.

I do, but I didn't before Portal. I hated science, but I liked Mecha. Now I know lots of things ...

Key questions

Let us revisit the quote from Lev Vygotsky at the end of the chapter: 'The invention of new methods that are adequate to the new ways in which problems are posed requires far more than a simple modification of previously accepted methods' (Vygotsky 1978: 58).

1   So, do computer games offer unique teaching and learning opportunities? Or do they merely cater for pupils already hooked on an 'edutainment' approach to their learning?
2   What might be some of the ethical issues that might arise from using digital games in the science classroom?
3   What advantages (and/or disadvantages) might commercial games offer over more specially created 'educational' games?

Further reading

Gee, J. P. (2007) *Good video games and good learning: Collected essays on video games, learning, and literacy.* New York: Peter Lang. In this seminal text, Gee explores a range of debates surrounding the use of computer games in the classroom. Although debated from an American perspective, many of his ideas and observations are transferable to the curriculum in England, Wales and Northern Ireland.

Miller, D. J., and Robertson, D. P. (2010) Using a games-console in the primary classroom: Effects of 'Brain Training' programme on computation and self-esteem. *British Journal of Educational Technology*, 41(2): 242–55. This is an interesting paper from two of the leading exponents of game-based learning in the United Kingdom. In it, the authors explore the use of a popular commercial game amongst primary school children and offer some important empirical data in terms of their attainment.

Crowe, N., and Flynn, S. (2013) Hunting down the Monster: Using Multi-Play Digital Games and On-Line Virtual Worlds in Secondary School teaching. In M. Leask and N. Pachler (eds) *Learning to teach using ICT in the secondary school* (3rd edn). Abingdon, Oxon: Routledge/Taylor & Francis. In this piece, the authors extend some of the debates begun in this chapter and offer some further examples of digital game based learning.

References

Barab, S. A., Arici, A., and Jackson, C. (2005) Eat your vegetables and do your homework: A design based investigation of enjoyment and meaning in learning. *Educational Technology*, 45(1): 15–20.

Bettelheim, B. (1976) *The ues of enchantment: The meaning and importance of fairy tales.* London: Thames & Hudson.

Brown, G. (2009) The Romanes Lecture. Given at the Sheldonian Theatre, Oxford, 27 February. Available online at: www.number10.gov/Page18472 (accessed November 2012).

Butt, S., Clery, E., Abeywardana, V., and Phillips, M. (2010) *Wellcome Trust Monitor 1.* London: Wellcome Trust. Available online at: http://www.wellcome.ac.uk/stellent/groups/corporatesite/@msh_grants/documents/web_document/wtx058862.pdf (accessed 12 December 2012).

Byron. T. (2010) *Do we have safer children in a digital world? A review of progress since the 2008 Byron Review.* Nottingham: DCSF Publications.

Crawford, C. (1982) *The art of computer game design.* Available online at: http://pdf.textfiles.com/books/cgd-crawford.pdf (accessed 2 November 2012).

Crowe, N. (2011) 'It's like my life but more, and better!': Playing with the Cathaby Shark Girls: MMORPGs, young people and fantasy-based social play. *International Journal of Adolescence and Youth*, 16: 201–23.

Crowe, N., and Bradford, S. (2006) Hanging out in Runescape. *Children's Geographies*, 4(3): 331–46.

Crowe, N., and Watts, M. (2012) 'When I click 'ok' I become Sassy – I become a girl': Young people and gender identity: subverting the 'body' in massively multiplayer on-line role playing games. *International Journal of Adolescence and Youth.* Available online at: http://www.tandfonline.com/doi/pdf/10.1080/02673843/2012.736868 (25 June 2013).

Crowe, N., and Flynn S. (2013) Hunting down the monster: Using multi-play digital games and on-line virtual worlds in secondary school teaching. In M. Leask and N. Pachler (eds) *Learning to teach using ICT in the secondary school* (3rd edn). Abingdon, Oxon: Routledge/Taylor & Francis.

Csikszentmihalyi, M. (1990) *Flow: The psychology of optimal experience.* New York: Harper & Row.

Dalton, J. P. (2000) *Online training needs a new course: The Forrester report.* Available online at: http://www.forrester.com/ER/Research/Report/Excerpt/ (accessed 20 March 2013).

Davies, B. (2005) Youth work: A manifesto for our times. *Youth and Policy* (National Youth Agency), 88(1): 23.

de Freitas, S. (2006) *Learning in immersive worlds: A review of game-based learning.* JISC E-Learning Programme.

DeWit, J., and Osborne, J. (2008) Engaging students with science: In their own words. *School Science Review*, 90(331): 109–16.

Dickey, M. D. (2005) Engaging by design: How engagement strategies in popular computer and video games and inform instructional design. *Educational Technology Research and Development*, 53: 67–83.

diSessa, A. A. (2000) *Changing minds: Computers, learning, and literacy.* Cambridge, MA: MIT Press.

Fullick, P. (2004) Using the internet in school science. In R. Barton (ed.) *Teaching secondary science with ICT*. Buckingham: Open University Press (pp. 71–86).

Freire, P. (1985) *The Politics of Education*. Massachusetts: Bergin & Garvey.

Gee, J. P. (2003) *What video games have to teach us about learning and literacy*. New York: Palgrave Macmillan.

Gee, J. P. (2007) *Good video games and good learning: Collected essays on video games, learning, and literacy*. New York: Peter Lang.

Gee, J. P., and Hayes, E. (2009) *Public pedagogy through video games*. Available online at: http://www.gamebasedlearning.org.uk/content/view/59/ (accessed 3 February 2011).

Gee, J. P., and Shaffer, D. W. (2010) *Looking where the light is bad: Video games and the future of assessment*. Epistemic Games Group Working Paper No. 2010–02. Madison: University of Wisconsin–Madison. Available online at: http://epistemicgames.org/eg/looking-where-the-light-is-bad/ (accessed 15 March 2012).

Groff, J., Howells, C., and Cranmer, S. (2010) *The impact of console games in the classroom: Evidence from schools in Scotland*. UK: Futurelab.

Hoskins, B., and Fredriksson, U. (2008) *Learning to Learn: What is it and can it be measured?* JRC, European Commission Document.

Ito, M. (2009) *Engineering play: A cultural history of children's software*. Cambridge, MA: MIT Press.

Jalongo, M. R. (2007) Beyond benchmarks and scores: Reasserting the role of motivation and interest in children's academic achievement. *Association for Childhood Education International*, 83(6): 395–407.

Jenkins, E. W. (1999) School science, citizenship and the public understanding of science. *International Journal of Science Education*, 21: 703–10.

Ke, F. (2009) A qualitative meta-analysis of computer games as learning tools. In R. E. Furdig (ed.) *Handbook of research on effective electronic gaming in education*. New York: IGI Global (pp. 1–32).

Ketelhut, D. J., Dede, C., Clarke, J., and Nelson, B. (2006) *A multi-user virtual environment for building higher order inquiry skills in science*. Paper presented at the 2006 AERA Annual Meeting, San Francisco, CA. Available online at: http://muve.gse.harvard.edu/rivercityproject/documents/rivercitysympinq1.pdf (accessed 1 January 2013).

King, K. (2001) *Technology, science teaching and literacy: A century of growth*. New York: Kluwer Academic/Plenum Publishers.

Kirkpatrick, D. L., and Kirkpatrick, J. D. (2006) *Evaluating training programs* (3rd edn). San Francisco, CA: Berrett-Koehler.

Klopfer, E., Osterweil, S., and Salen, K. (2009) *Moving learning games forward*. Cambridge, MA: The Education Arcade.

Larson Mclarty, K., Orr, A., Frey, P., Dolan, R., Vassileva, V., and McVay, A. (2012) *A literature review of gaming in education*. Pearson. Available online at: http://www.pearsonassessments.com/pai/ai/research/researchandresources.htm (accessed 12 December 2012).

Laugksch, R. C. (2000) Scientific literacy: A conceptual overview. *Science Education*, 84: 71–94.

Leite, L., Vieira, P., Silva, R., and Neves, T. (2007) The role of WebQuests in science education for citizenship. *Interactive Educational Multimedia*, 15: 18–36.

Livingstone, S., and Bober, M. (2005) *UK children go online: Final report of key project findings.* London: Economic and Social Research Council.

McPherson, A. (1969) Swing from science or retreat from reason? *Universities Quarterly*, 24: 29–43.

Mezirow, J. (2000) *Learning as transformation: Critical perspectives on a theory in progress.* San Francisco: Jossey Bass.

Morrow, V. (2005) *Ethical research with children.* London: Open University Press.

Oblinger, D. (2004) The next generation of educational engagement. *Journal of Interactive Media in Education*, 2004(8): 1–18.

OECD (1999) *Measuring student knowledge and skills: A new framework for assessment.* France: OECD Publications.

OFCOM (2011) *Children and parents: Media use and attitudes report.* Available online at: http://stakeholders.ofcom.org.uk/binaries/research/media-literacy/oct2011/Children_and_parents.pdf (accessed 14 November 2012).

Osborne, J. F., and Collins, S. (2001) Pupils' views of the role and value of the science curriculum: A focus-group study. *International Journal of Science Education*, 23(5): 441–68.

Owen, S., Dickson, D., Stanisstreet, M., and Boyes, E. (2008) Teaching physics: Students' attitudes towards different learning activities. *Research in Science and Technological Education*, 26(2): 113–28.

Parker, K. (2004) *Free play: The politics of the video game.* Reason Online. Available online at: http://www.reason.com/0404/fe.kp.free.shtml (accessed 6 June 2011).

Poole, S. (2000) *Trigger happy: The inner life of video games.* London: Fourth Estate.

Postman, N. (1996) *The disappearance of childhood.* London: Vintage.

Prensky, M. (2001) Digital natives, digital immigrants. *On the Horizon* (MCB University Press), 9(5).

Pricewaterhouse Coopers (2010) *Social multi-player gamers named as UK technology's hottest prospect.* Press release 19 October. Available online at: http://www.ukmediacentre.pwc.com/News-Releases/Social-multi-player-gamers-named-as-UK-technology-s-hottest-prospect-f46.aspx (accessed 24 October 2010).

Raine, L., and Horrigan, J. (2005) *A decade of adoption: How the internet has woven itself into American family life.* Washington, DC: PEW, Internet and Family Life.

Rieber, L. P. (1996) Seriously considering play: Designing interactive learning environments based on the blending of microworlds, simulations, and games. *Educational Technology Research and Development*, 44(2): 43–58.

Roth, W., and Désautels, J. (2004) Educating for citizenship: Reappraising the role of science education. *Canadian Journal of Science, Mathematics and Technology Education*, 4(2): 149–68.

Rutherford, F. J., and Ahlgren, A. (1989) *Science for all Americans.* New York: Oxford University Press.

Shute, V. J., Ventura, M., Bauer, M. I., and Zapata-Rivera, D. (2009) Melding the power of serious games and embedded assessment to monitor and foster learning: Flow and grow. In U. Ritterfeld, M. Cody, and P. Vorderer (eds) *Serious games: Mechanisms and effects.* Mahwah, NJ: Routledge, Taylor & Francis (pp. 295–321).

Smith, E. (2005) *Analysing underachievement in schools.* London: Continuum.

Smith, E. (2011) Staying in the science stream: Patterns of participation in A-level science subjects in the UK. *Educational Studies*, 37(1): 59–71.

Squire, K. (2006) From content to context: Video games and designed experiences. *Educational Researcher*, 35(8): 19–29.

Squire, K., and Patterson, N. (2010) *Games and simulations in informal science education*. WCER Working Paper No. 2010-14. Madison: University of Wisconsin–Madison, Wisconsin Center for Education Research. Available online at: http://www.wcer.wisc.edu/publications/workingPapers/papers.php (accessed 12 November 2012).

Toplis, R. (2011) *How science works: Exploring effective pedagogy and practice.* London: Routledge.

Vygotsky, L. (1933) Play and its role in the mental development of the child. Source: *Voprosy psikhologii*, 1966(6); translated: Catherine Mulholland; transcription/markup: Nate Schmolze; online version: Psychology and Marxism Internet Archive (marxists.org) 2002. Available online at: http://www.marxists.org/archive/vygotsky/works/1933/play.htm (accessed 3 December 2011).

Vygotsky, L. (1978) Interaction between learning and development. In M. Cole, V. John-Steiner, S. Scribner, and E. Souberman (eds) *Mind in society: The development of higher psychological processes.* Cambridge, MA: Harvard University Press.

Wellcome Trust (2011) *Exploring young people's views on science education.* London: Wellcome Trust.

Chapter 8

Language and communication in the science classroom

Susan Rodrigues

Introduction

Recently, my friends Ann and John were having a conversation about the gender symbols (a circle with a cross on top and a circle with an arrow at an angle). Frustratingly, at that particular moment in time we could not remember which was associated with males and which related to females. We tried to dredge up clues to help us but as this conversation took place out of context, having just finished with another unrelated conversation, we struggled to assign symbols to genders. In some ways, the difficulty we experienced is similar to the difficulties pupils experience when they attend science lessons. They vaguely recall the symbols and language used in their previous science lesson, they have usually just emerged from another class (using other practices, other symbols and codes) and are expected to make a speedy transition to take into account a different teacher's style (in terms of conduct and pedagogy). The last thing these pupils need is an additional barrier introduced by convoluted language, in terms of concept, content or context.

The word 'quark' was coined by the physicist Murray Gell-Mann, who said it was a nonsense word in James Joyce's *Finnegans Wake* (1939). Unfortunately, for many pupils, their science lessons appear to be littered with nonsense words, phrases and meanings. This may be because the pupils find the science words to be unusual, and it may also be because the science words are contextualised inappropriately. Not only are pupils expected to make sense of unfamiliar science words, but far too often they find themselves having to navigate complex communication pathways in their science lessons. These communication pathways become complicated in science lessons because teaching science occurs without considering the importance of vocabulary acquisition, translating skills and context. We forget what it was like before we had established our level of understanding, our command of the language and our engagement with the subject for personal interest's sake. How we, as teachers, address this issue of communication rather depends on our model of science education.

If we hold a deficit model of science education, then the fall back position is that the pupil lacks the necessary knowledge regarding science, and as teachers it is our role to communicate this knowledge much in the way of a postal delivery: drop off a parcel, get a signature to acknowledge receipt of delivery and consider the task fulfilled. Knowledge delivered. If, on the other hand we have a contextual model of science education, then dialogue between the teacher and pupils becomes a central focus. However, it does not have to be an either/or situation, for neither the deficit nor the contextual model stand alone. There is a need to, on occasion, transmit and receive information, just as there is on occasion a need to consult and collaborate. In both situations, the way in which language is used is crucial. It is important that coherent and cohesive language is used, in order to see plain speaking in science lessons. Plain speaking in science classrooms matters for several reasons. Two of the reasons are:

- *From a cognitive science perspective* – It is often suggested that structure is applied to concepts and knowledge when stored in memory. A richer structure enables better access to this stored information. A richer structure is more likely if pupils are not bedazzled or confused by the language used.
- *From a socio-cultural research perspective* – It has been suggested that communities construct knowledge by sharing information and understanding. So the role of language is key to if we want pupils to become informed members of the science community.

Not surprisingly then, there is a fundamental need to consider the place of communication in the establishment of the negotiation of meaning in science lessons.

In this chapter I discuss the language in science classrooms and attempt to make the case for plain speaking in both the nature and role of language. I consider three main aspects in this chapter: patterns of talk; the nature of language; and the role of language in science lessons. I also reflect, albeit briefly, on the influence of assessment in science lessons with regard to comprehension and techno-babble language issues. In this chapter I hope to show how language is often used in a way (albeit inadvertently) that leads to unnecessarily complicating science learning. I want to show how we lull ourselves into a false sense of security in our assessment of pupils' comprehension of science concepts. We tend to ask the pupils to repeat or use techno-babble, but they may have little comprehension for the words being uttered, written or heard. In this chapter I will show how the pattern of talk and types of engagement may generate barriers to the development of understanding. I will then show how the content of talk influences developing understanding, and I provide examples of how we test and record attainment and claim to show understanding has been achieved, when in reality that may not be the case.

Language and understanding

It could be argued that science lessons are predominantly about describing and modelling our world to help those in our classes to develop a more informed understanding of our world. In describing and modelling our world, we use language that is thought to have three meta functions. One of these meta functions is called ideational (a means of representing patterns of experience). These patterns of experience are generated through particular grammar of a language. The grammatical 'process' in language refers to the verb or verb group. Surrounding the verb or verb group are selections of grammatical participants (nouns) that participate in the process. Sometimes surrounding the participants we find grammatical circumstances (adverb, adverbial phrase or prepositional phrase) that elaborate the process or participant. The language pupils use (in terms of process, participants and circumstances) in school science and the language they use in their everyday life has significant differences. No one would argue with the fact that word selection, word meanings and their use reflects and constructs a context. The most obvious example being that while social language tends to be informal and interactive (Schleppegrell 2001), science lesson language warrants navigating various literacy techniques, dealing with embedded clauses as well as high lexical density (high number of technical terms per sentence), often sifting through significant amounts of abstract or technical language, and on many occasions never revisiting these new terms/ideas/literacies in any significant way. Not surprisingly, as the literature indicates, many pupils find it very difficult to cope with not just the language commonly encountered in school based science lessons (Unsworth 2000), but also experience difficulty in the way the language is used in these classroom settings. As a result, for many pupils, the language they encounter in science lessons renders the pupils helpless and frustrated.

Unfortunately, often this helplessness goes unrecognised, because the assessment deployed in science classrooms allow pupils to generate a false impression. The pupils may not understand the ideas/discussion/concepts in reality, as little understanding is present, but they can repeat the phrases required in response to questions that seek to determine recall rather than comprehension. More on this later.

Pattern of talk

The language seen in typical out-of-classroom conversation is more open, less prescriptive and more dialogic. In contrast, most school science lessons tend to involve what has been called an initiate-respond-follow (IRF) up pattern of talk (see the work of Barnes and Todd 1995; Rodrigues 2010; Sinclair and Coulthard 1975) or a monologue. Sinclair and Coulthard (1975) in their study of discourse identified common patterns of turn-taking, whereby pupils and teachers took turns to speak according to a perception of their roles. In most school science lessons, the words also tend to be more formal in their links and associations with neighbouring words (Lemke 1985).

Sometimes the IRF pattern of talk is viewed as a negative dynamic. It is criticised for its perceived imbalance in power between the pupil and the teacher. As a result it is often considered unfavourable, for there is an implied lack of learning opportunity afforded to the pupil within the IRF pattern. I would suggest that it is not so much a case of advocating a symmetrical relationship with regard to power, between the teacher and the pupil, because in reality the pupil is still the apprentice. Hence the IRF pattern should not be held accountable for poor learning opportunity. For while pupils may bring with them a body of knowledge and experience, in most cases this body of knowledge and experience is a work in progress.

If the IRF pattern of interaction adopted a 'power with' value, then there might well be more scope for interactivity in the form of meaningful response and follow-up within the IRF. In promoting meaningful response, the 'initiate' statement or question needs to allow for valued, informed response rather than simply afford an opportunity to check on the pupils' ability to recall techno-babble. The IRF needs to help the pupils construct, develop and refine their understanding and make the construction visible to their teacher.

Many strategies have been advocated in an attempt to change and challenge the pattern of talk commonly found in classrooms in order to help make pupils' thinking and understanding more visible. Researchers have identified the use of questioning, cognitive conflict and 'wait time' to re-address the issue of turn-taking in science lessons. More recently we have seen researchers support and promote argumentation that builds on the 1980s notion of cognitive conflict. The current case for argumentation is often made on the basis of: supporting communicative, reasoning and cognitive competences; encouraging pupils to talk and write science; and enculturation. In essence, argumentation looks for justification of knowledge and it sees argument as persuasion (Toulmin 1958). However, this is not enough. If we accept the importance of science discourse in the construction of science knowledge (Latour and Woolgar 1986) and the accept role of social interaction in learning in terms of higher order processes originating through language mediation (Vygtosky 1978, 1986), then justification and persuasion provide a start. If we accept the statement 'For many pupils the greatest obstacle in learning science – and also the most important achievement – is to learn its language' (Wellington and Osborne 2001: 3), then addressing the surface activity of shifting power in turn-taking (through argumentation or another similar strategy) provides limited help for pupils in terms of learning its language. For to learn 'its language' requires pupils to develop strategies that help them interpret the logic and grammar of 'scientific' ways. Learning 'its language' requires skills in navigating the grammar of science language, translating embedded clauses, dealing with the high lexical density they encounter, addressing abstract and technical language: becoming familiar with the context, concepts and content of science. Oyoo's (2004) review of pupils' understandings of science words indicated that pupils have to be able to recognise the words before they establish meanings for the words in the context of use.

It is worth noting that the meanings for words relates not just to what would normally be considered science words, but also includes teacher use of non-technical words. We need to consider the difficulties pupils encounter with contextual meanings for science and non-technical words when used in science classrooms because, if we do not, then language in science classrooms remains an undecipherable language for most pupils.

Content of talk

For many pupils their science lessons are littered with, in some cases, the one-off use of words, phrases and implications that make limited if any sense at the time, let alone later. Nevertheless the pupils are expected to use these processes, participants and circumstances to construct and convey their understanding of science during a period of assessment. In some cases the technical and non-technical words are used with little awareness of the difficulty they generate for the pupils.

Words like 'illustrate', 'discuss', 'describe' or 'define' are often in used in science lessons and may well be considered to be understood through osmosis and frequent use. In reality, pupils may experience difficulty in interpreting these 'process' words within the science lesson setting. There may be no shared explicit awareness of the functional value of these non-technical words in science lessons. For example, what makes a discussion different to a definition or a description? This, compounded with the challenge in using science terms (technical words), may generate a lack of understanding. Thus, as a teacher, it is important to provide explanations of non-technical words and concepts for science lesson purposes, as well as provide explanations for the science (technical words).

In many instances, explaining the science results in the use of what Johnstone (1991) labelled 'microscopic, macroscopic and symbolic language'. For teachers, moving between these three types of language is effortless. Their experience and familiarity with macroscopic, microscopic and symbolic language allow them to use the macroscopic, microscopic and symbolic language interchangeably. So they move between these strands of language without realising that the pupils in their class may be experiencing difficulty in making the transitions. Rodrigues and Bell (1995) showed that this posed problems for some pupils as they did not grasp the implications and significance of the macro, micro or symbolic representations.

In recent science education times, there have been calls to address abstract and technical language through the use of pupil-familiar contexts in order to enable pupils to engage in classroom dialogue. However, if we consider Halliday's notion of register in terms of field, mode and tenor, then we can see that simply changing the context from a less familiar one to a more familiar one does not necessarily equate with engagement, if the classroom language field (content), tenor (tone) and mode (way of communicating) remain the same. For a pupil to understand science discourse (be it spoken or written), that discourse has to have coherence and cohesion. It has to show continuity of topic and provide reference.

Cohesion occurs when the interpretation of a particular element within the discourse is dependent on that of another element (Halliday and Hasan 1989) and this generates continuity. Coherence tries to ensure that talk is more than just a string of words. If we said, 'Susan likes Chemistry. She sang in Paris.' The two sentences are cohesive. ('Susan' can be linked to the sentence that followed through the use of 'She'. This provides a cohesive link.) However, the sentences are not coherent (liking science and singing in Paris do not follow). Yet, how often during a science lesson do we as teachers merge talk signalling direction with talk providing science content, and expect the pupils to see the discourse as both cohesive and coherent (Hasan 1984)?

From a linguistics perspective, cohesion is determined by taking into consideration the links that relate clauses and sentences. If these links are weak, a pupil may experience difficulty in trying to figure out the teacher's idea or view. In most cases, analysis of casual conversation involving familiar topics shows that there are fewer cohesive links. This is because those doing the talking are familiar with each other and the topic. As a result, one talker infers what the other talker means with little need to include cohesive links. Unfortunately, as Rodrigues and Thompson (2001) showed, in a science classroom setting, fewer cohesive links do not work, because one talker (be it the teacher or pupil) tends not to be as successful at inferring what the other talker means. Hence, in science classrooms, having fewer cohesive links affects what is meant and results in confusion.

In science lessons, if the cohesive ties are insufficient and if the technical and non-technical language remains vague, then neither the teacher nor the pupil is successful in inferring appropriate meanings. Thus if a science teacher's talk has weak links during the science lesson, then the pupils may experience difficulty in ascertaining the teacher's intention. The pupils then start to rely on inferences, not all of which may be applicable or appropriate. Likewise, if the pupils in a science lesson provide a response with weak links, then the teacher has to rely on inference that, again, may result in further confusion. For example, if during the course of a chemistry lesson the teacher talks about 'a ground state electron', the chemistry teacher is talking about that electron having a particular amount of energy. However, the pupil has to infer this. Unfortunately they may not share the interpretation of the word 'state' in terms of energy state. The pupils may simply see the notion of 'state' as a spatial entity. The pupil needs assistance in understanding that phrase: in this case, the use of the word 'state' in terms of an electron having specific energy.

However, just as too few links cause confusion, too many links may result in the talk becoming pedantic and difficult to sift through. There is a need to ensure that talk in science classrooms finds the right balance: links that are not dense, but are sufficiently cohesive to enable the listener (either the teacher or the pupil) to see the talk as coherent. If they are to make sense of the contribution of the other conversation participant, the talk needs to be coherent and cohesive.

Teaching science without deploying the skills needed to support what could be considered second language acquisition further complicates science in schools, or, as Edwards and Westgate (1994: 20) wrote, 'Registers associated with particular school subjects can be seen as containing both the terms embodying concepts essential to their practice, and markers of academic boundaries between one subject and another, and between school knowledge.' Science classrooms rarely take into consideration the importance of vocabulary acquisition (the terms embodying concepts), translating skills (the markers of academic boundaries) and context (between the subjects and between school knowledge). We may teach school science as the history of science rather than science as currently experienced, but even then we tend not to explain to the pupils the etymology or history of the words we use, thus leaving the words to emerge and remain in the form of convoluted techno-babble.

Why in science does a definition often require a reader to seek the definition of words used in the original definition? Take for example the definition of 'capillary action':

> The drawing of a liquid up the inside of a small-bore tube when adhesive forces exceed cohesive forces, or the depression of the surface of the liquid when cohesive forces exceed the adhesive forces.

In plain language, what does this actually mean? What do adhesive and cohesive forces mean?

Many years ago I used to teach chemistry in a secondary school. I remember telling pupils that 'reduction is the gain of electrons'. With hindsight I imagine that for many pupils that sentence was a nonsense, for the word 'reduction' implies lessening or decrease, and I was defining it with the word 'gain'. From a language grammar perspective, the sentence would appear to the pupils to contain two process words (verbs: to reduce, to gain) that appear to contradict each other. The issue arises because, in my chemistry lesson, the word reduction was not being used as a process (verb) word in the language framework, but was being used as a participant (noun) in the science framework. It might have helped pupils if the framework of the sentence was explained in terms of the history of developments in science, in particular with regard to the concept of oxidation and reduction. If the history of developments were explained, showing how the definition came to be refined and be used as a noun, while indicating an awareness of the contradiction in language use in science and in everyday use, then pupils may not view this as something that does not make sense. It may help pupils build better and richer structures when storing the information.

The previous example also highlights the fact that words used in science subjects may be used to mean different things in everyday language (or indeed have different meanings within the different sciences). This may be as previously illustrated, because the words are used differently in terms of process, participant and circumstance, or it may be because they are simply defined differently. This issue is not new. Vygotsky (1986) discussed the notion of 'everyday' and 'scientific' concepts. Halliday (1993) discussed reconstructing 'commonsense' (everyday) language into an abstract form as 'educational' knowledge to produce 'scientific' language. To do this, Halliday (1993) suggested that pupils need a particular 'grammar' if they are to reconstruct the everyday language to negotiate meaning and form more complex kinds of understanding for science lessons.

Science, we know, uses existing words (for example, reduction) and new words (for example, quarks) to explain, to predict, to challenge and to confirm scientific ideas. Not content to use everyday words in science to mean different things, we also use the same words to mean different things within the different science disciplines. Take, for example, 'resonance':

- In chemistry, resonance can be defined as a way of describing delocalised electrons – the concept in which two or more equivalent dot formulas for the same arrangement of atoms (resonance structures) are necessary to describe the bonding in a molecule or ion.
- In physics, resonance can be defined as the increase in amplitude of oscillation of an electric or mechanical system exposed to a periodic force whose frequency is equal or very close to the natural undamped frequency of the system.

Not content with using complex terms or the same words to mean different things, we have also shown that we use everyday words to generate counter-intuitive meanings. For example, consider the paragraph below, once again on the topic of oxidation and reduction.

> Oxidation state shows the total number of electrons that have been removed from an element (a positive oxidation state) or added to an element (a negative oxidation state) to get to its present state. Every time you oxidise an element by removing another electron from it, its oxidation state increases by 1.

For many pupils the paragraph above is challenging. The pupils have to equate words in a counter-intuitive way in order to understand the concept of oxidation states. Removing something makes it positive or results in an increase, while adding something makes it negative.

Second language learning often sees the context of new words as key to developing and understanding the word. In contrast, in science lessons, the nature and context of the word are largely ignored. The Latin word 'formula' describes a contract, regulation or rule, and in the 1630s it represented words used in a ritual. The scientific use of 'formula' in chemistry is ascribed to Carlyle who, in 1837, used it to describe a rule slavishly followed without understanding. While Newton is often associated with the term 'force', it is a word that was used in the 1300s. The term 'mass' probably comes from French word 'masse' (pile, crowd, ingot), the Latin word 'massa' and the Greek word 'maza' (kneaded dough, barley cake). It was used in the late fourteenth century and only became a scientific term three hundred years later, around 1704. In some respects, providing some of this background information to explain the terms and language used in science provides a context for the word, and may help the pupils to establish better internalised networks. As such, it may help the pupils with their knowledge construction, for, as cognitive science has suggested, the richer the structure, the better the access to stored knowledge.

Currently topical, given the regular updates from CERN, various words (such as quark and the 'God particle', to mention but two) have found themselves in the public domain via the mainstream media. If, as seen in a recently-published edition of a free UK-wide paper, a journalist can present a description of the importance of this 'God particle' using plain speak, without diminishing the power of the words used by the scientists, why, as teachers, are we not adopting a similar approach?

I am not advocating that we do not teach pupils the scientific words. I am suggesting that the way in which these words are introduced and taught needs to be reconsidered. Techno-babble is rendered babble because it is not understood. For many pupils, the words used in science education are far from educational, and remain nonsensical or just babble. It has been well-established that many pupils find it very difficult to deal with language encountered in school science lessons and school science tasks. As mentioned above, this often results in science language rendering pupils helpless and frustrated.

The role of language

Science lessons try to re-present and re-context professional science discourse, or as Christie (1998: 52) wrote, it construes what is often 'not familiarly available'. According to Martin (2003), pupils feel they need to know this information for examinations, so they memorise it as facts and recite it, often without understanding that the scientists have argued for, selected, interpreted and packaged this information to enable it to have a shared understanding within the science community. Unfortunately, instead of verifying student understanding of this information, we often rely on their ability to recall this information as proof of their understanding.

Thus, in reality, we may be failing to gauge student understanding because we frame questions in such a way as to make the understanding requirement relatively negligible. In many instances, the purpose of language used in science lessons does little more than provide an opportunity to test for recall at a later date. To illustrate this, read the following text and answer the questions that follow.

In a gaisel engine, air is compressed isomatically with an inflexion ratio between 15 and 20. This inflexion raises the temperature to the ignition temperature of the fuel mixture formed by injecting fuel once the air is inflexed. The ideal air-cycle is based on a reversible isomatic inflexion followed by a constant pressure combustion process, then an isomatic expansion as a power stroke and an adometric exhaust. Since the inflexion and power strokes of this idealised cycle are isomatic, the efficiency can be deduced from the constant pressure and constant volume processes.

- How is air compressed in the new gaisel engine?
- What happens once the air is inflexed?
- Summarise the standard air cycle model.

The three questions/statements can be answered without any real understanding of the paragraph on this fictional gaisel engine. The third statement could be classified as a relatively higher order task, as it requires the reader to take information and represent it as a summary. However, in reality, it does not warrant any measure of comprehension, just literacy skills in terms of finding the key sentences to represent in answering the question. Some may read the paragraph on the gaisel engine and say that the words used are a nonsense and do not exist: gaisel, isomatically, inflexion etc. True. But the gaisel engine paragraph is based on a paragraph describing the diesel engine in which key words were replaced. So, for example, in the text presented earlier, 'adiabatically' was replaced with 'isomatically'. Pupils may well consider both to be nonsense words.

Often, to overcome the challenge in interpreting abstract terms (or what pupils consider nonsense words), we resort to the use of analogies. However, we need to proceed with caution, for analogies require the pupils to have an understanding of the field. We also fail to recognise that pupils sometimes overextend the analogy or focus on superficial similarities resulting in further confusion. Thus analogies need to be considered carefully before being used in science lessons.

As seen earlier, the nature of the language pupils need to use in school science and the language they need to use in their everyday life have significant differences. But these differences include more than the nature of the language: they also include the role of language. Few would argue with the fact that word choice, their meanings and the way in which we structure their use reflects and constructs our setting and our understanding of our setting. Science lessons promote the repetition of word use, rather than an understanding of the word in itself, or within a given context. In many respects it is akin to teaching pupils the 8-times table which they are then able to recite, without them having any real understanding of the concepts involved.

Conclusion

The language in science subjects may differ from everyday social language used by students. For example, texts may become increasingly abstract and dense, and they may also involve discursive conventions that are often subject-defined. Yet, to a certain extent, these challenges and difficulties are ignored. We may attend to helping pupils recognise words, and we may help them acquire vocabulary and gather information, but we tend not to consider the way recognition, acquisition and use of these words reflect or reduce comprehension. Do the pupils understand the language they are using? Asking them to use the words and recall the vocabulary is not an indication of comprehension.

If we were to start by looking at the language we use in science lessons with a view to promoting plain speaking, then first we need to look at our grammar, in terms of process, participant and circumstance. If the word is being used in everyday language as a verb but is used in science as a noun, then as teachers we need to make that explicit. If the word appears to contradict everyday use or interpretation, then through the use of plain language we need to provide the historical context for specific word-use in science. We ought to have the skills required to provide explanations in sentences that do not involve embedded clauses that warrant further definition. We also need to consider the coherence and cohesiveness of our language. Make sure the links that tie our statements are obvious. If we fail to do this, then school-acquired language for science subjects may appear to show students are learning science, when in reality little understanding is present.

As science communicators, we need to be better at understanding language practices, symbols and codes, so as to limit the development of convoluted language. Language is functional. I accept that different ways of using words in science, in out-of-school activity and in personal lives generates different kinds of doing and different kinds of outcomes. However, that does not mean that confusion should be allowed to thrive in science lessons simply because we are lazy with our language. Chunking our explanations into smaller steps and providing insight into technical and non-technical terms may help pupils understand the key ideas. Ensuring cohesion and coherence is present, by becoming more aware of the lexical density in our dialogue with pupils, may help pupils comprehend science. Making the links explicit in order to reveal content in terms of process, participant and circumstance may also help pupils make better sense of science concepts. If we want to assist pupils in becoming members of the everyday and career science communities, then we have to help them acquire the language to communicate within those communities.

Key questions

1   Teachers have a good knowledge of curriculum content and understanding of the issues that are likely to confuse children or challenge their thinking. But do your questions match the instructional purpose? Do you 'chain' the

question and answers, so that ideas are developed or changed? Some questions will invoke a range of responses and encourage divergent thinking, others will require only single word responses.

2   Teachers encourage learning talk through activities that require children to respond, but do they do so in extended utterances? How do you model language that goes beyond what learners are able to produce alone? Do you really listen and respond to the content of students' utterances, challenging, probing and extending their meanings?

3   When learners initiate dialogue, do you 'withdraw from the floor' to allow them the space to talk? How well do you get students to address the whole class in an intelligible and articulate way, and to listen carefully to each other's contributions?

Further reading

Alexander, R. J. (2004) *Towards Dialogic Teaching: Rethinking classroom talk*. York: Dialogos. This is an excellent monograph about ways to engage students in talking and learning.

Lemke, J. L. (2003) *Talking Science: Language, learning, and values*. Norwood, NJ: Ablex. A really good book about 'talking science' and the ways in which classroom teachers do (and should) engage learners in discussion.

References

Barnes, D., and Todd, F. (1995) *Communication and Learning Revisited: Making meaning through talk*. Portsmouth, NH: Boynton/Cook Publishers Heinemann.

Christie, F. (ed.) (1998) *Pedagogy and the Shaping of Consciousness: Linguistic and social processes*. London: Cassell.

Edwards, A. D., and Westgate, D. P. G. (1994) *Investigating Classroom Talk* (2nd edn). London: The Falmer Press.

Halliday, M. A. K. (1993) Towards a language-based theory of learning. *Linguistics and Education*, 5: 93–116.

Halliday, M. A. K., and Hasan, R. (1989) *Language, Context, and Text: Aspects of language in a social-semiotic perspective*. Oxford, UK: Oxford University Press.

Hasan, R. (1984) Coherence and cohesive harmony. In J. Flood (ed.) *Understanding Reading Comprehension*. Newark, Delaware: International Reading Association.

Johnstone, A. H. (1991) Why is science difficult to learn? Things are seldom what they seem. *Journal of Computer Assisted Learning*, 7(2): 75–83.

Latour, B., and Woolgar, S. (1986) *Laboratory Life: The construction of scientific facts*. Princeton, NJ: Princeton University Press.

Lemke, J. (1985) *Using Language in the Classroom*. Victoria, Australia: Deakin University Press.

Martin, A. J. (2003) *How to Motivate Your Child for School and Beyond*. Sydney: Bantam.

Oyoo, S. O. (2004) *Effective Teaching of Science: The impact of physics teachers' classroom language*. Unpublished PhD Thesis, Faculty of Education, Monash University, Australia.

Rodrigues, S. (2010) Exploring talk: Identifying register, coherence and cohesion. In *Using analytical frameworks for classroom research: Collecting data and analysing narrative.* New York: Routledge.

Rodrigues, S., and Bell, B. (1995) Chemically speaking: Students' talk during chemistry lessons. *International Journal of Science Education*, 17(6): 797–809.

Rodrigues, S., and Thompson, I. (2001) Cohesion in science lesson discourse: Clarity, relevance and sufficient information. *International Journal of Science Education*, 23(9): 929–40.

Schleppegrell, M. (2001) Linguistic features of the language of schooling. *Linguistics and Education*, 12: 431–59.

Sinclair, J., and Coulthard, M. (1975) *Toward an Analysis of Discourse: The English used by teachers and pupils.* Oxford: University Press.

Toulmin, S. (1958) *The Uses of Argument.* Cambridge: Cambridge University Press.

Unsworth, L. (2000) *Researching Language in Schools and Communities: Functional linguistic perspectives.* London: Continuum.

Vygotsky, L. S. (1978) *Mind in Society. The development of higher psychological processes.* Cambridge, MA: Harvard University Press.

Vygotsky, L. S. (1986) *Thought and Language.* Cambridge, MA: MIT Press.

Wellington, J., and Osborne, J. (2001) *Language and Literacy in Science Education.* Philadelphia: Open University Press.

Chapter 9

Questions and inquiry in science: 'no questions, no learning'?

Helena Pedrosa-de-Jesus and Sara Leite

Introduction

> Unfortunately, most students ask virtually none of thought-stimulating types of questions. They tend to stick to dead questions like 'Is this going to be on the test?', questions that imply the desire not to think. Most teachers in turn are not themselves generators of questions and answers of their own, that is, are not seriously engaged in thinking through or rethinking through their own subjects. Rather, they are purveyors of the questions and answers of others – usually those of a textbook (...) no questions equals no understanding. Superficial questions equal superficial understanding. Most students typically have no questions. They not only sit in silence; their minds are silent as well.
>
> (Paul and Elder 2007)

Is that the case? Is it true that 'no questions equals no understanding'? In this chapter we challenge that view – but only in part: our perspective is not that a lack of questions means a lack of understanding, rather that 'many more questions equals much more understanding'. That is, we do not begin from a zero start-point, but we certainly do see the value in a multitude of thought-stimulating questions – some, of course, from the teacher, but mainly by the learners: we want students' minds to be noisy, not silent.

This chapter falls into two parts. In the first, we discuss the importance of classroom questions, where it is the student who asks the questions. This can be a process that is difficult for teachers to manage – after all there are potentially thirty questioners in the room and usually only one teacher-responder. Moreover, as Paul and Elder suggest above, questions 'from the floor' can challenge science teachers (all teachers) to engage in thinking and re-thinking through their own subject knowledge. However, we are fully convinced of the virtues of student questioning and explore here some of the reasons why.

In the second part of the chapter, we describe how one of us (Leite) fostered the participation of students in her class, collecting their ideas and queries and reflecting on the information collected, with the intention of making changes in her planning for the next tranche of lessons. After each lesson, she designed flexible instruction that took account of the students' questions she had collected, and these lesson plans were then implemented through the next series of classes. So, student responses during 'question moments' were again collected, and then a further stage of re-design took place – and so on, as the series of lessons progressed. Our discussions chart the strategies (or teaching tools) we used, the subject matter and the students involved. This is essentially a co-constructivist approach to teaching, where knowledge and understanding is built and then 'orchestrated' (Watts and Pedrosa-de-Jesus 2005) within a classroom community.

Why ask questions?

> Most people think that genius is the primary determinant of intellectual achievement. Yet three of the all-time greatest thinkers had in common, not inexplicable genius, but a questioning mind. Their intellectual skills and inquisitive drive embodied the essence of critical thinking. Through skilled deep and persistent questioning they redesigned our view of the physical world and the universe. When we consider the work of these three thinkers, Einstein, Darwin, and Newton, we find, not the unfathomable, genius mind. Rather we find thinkers who placed deep and fundamental questions at the heart of their work and pursued them passionately. Would that we had students who did the same.
>
> (Paul and Elder 2007)

While we may not in the business of teaching geniuses, that last sentence is telling: 'Oh that we had students who ask deep and fundamental questions'. Not all teachers would welcome that state of affairs: it would be decidedly hard work to field a persistent stream of deep-seated questions in every lesson of every day. Perhaps it is simply easier to lecture to a sea of blank faces regardless of whether they show signs of life or not. Our chosen route is undoubtedly difficult: we are committed to making school science a lively and inquisitive business. The position we adopt in this chapter is that asking questions is an essential human quality. As we noted above, we shape our discussion from within a constructivist perspective and, in this, we side with Dabrowska and Lieven (2005), for whom question-asking is an act of personal meaning-making; something that has been said in several of the other chapters in this book. Question-askers, they say, build questions by recycling and recombining previously-experienced 'chunks' of language, knowledge and understanding. We recognise that such student-centred approaches are commonly challenging for teachers, not least because they 'require teachers to assume a guiding role and to simultaneously attend to many different aspects of the classroom' (Brush and Saye 2000: 8). The approaches we advocate (and discuss in more detail later) require us to understand students' points of view about the matters discussed in classrooms. In order to do this, on the one hand, we must adopt a reflective attitude towards the interpretation of classroom activities, and, on the other, an investigative attitude to collecting questions and answers. In our view, there is also a third, forward-looking requirement: the teacher must also develop and implement tools and strategies fitted to improving the process of teaching and learning in future lessons.

There is long history of psychological and educational research that sees questioning as a fundamental process within learning – we are not alone in seeing the educational benefits of students asking questions. There are numerous studies (e.g. AAAS 1989, 1993; Dori and Herscovitz 1999; Cuccio-Schirripa and Steiner 2000; Chin *et al.* 2002; Harper *et al.* 2003) that show that, when students ask questions, they are:

- connecting and clarifying information into memory;
- improving reading comprehension and content retention;
- using an important meaning-making and learning skill;
- using higher ordered and creative thinking;
- improving their problem solving skills;
- developing the ability to identify anomalies in their own understanding of the topic;
- providing the teacher and classmates with a window into his or her thinking and evidence of the conceptual change;
- taking steps towards a lifetime of self-directed learning;
- more authentically following the practices of science – a discipline that is based on asking questions and finding answers.

The key issue here is that many questioners do not ask questions simply for information, but do so because they want to understand better. But Paul and Elder (2007) above do have a point: most learners do not actually ask their questions. In their work, Good *et al.* (1987) discuss this notion of 'passivity', and suggest that there are at least two classroom pressures at work. Llow achieving students have learned to become less involved in classwork, to keep quiet and stay 'below the radar', to be non-question-askers. On the other hand, high achievers are sensitive to the possible cost of asking question – the embarrassment of asking what appears to be stupid question (van der Meij 1994). The end result can be that no one will ask, and the room remains quiet. There are other reasons.

Why *not* ask a question?

At a surface level, it would seem rather straightforward to ask students about their uncertainties and difficulties. However, as Maskill and Pedrosa-de-Jesus (1997) have pointed out, pupils do not feel at ease when answering teachers' questions, because they link questioning to assessment, and perhaps from a natural their fear that exposing ideas might reveal learning problems. In Pedrosa-de-Jesus and Watts (2012) we explore how asking a question in class gives rise to feelings of exposure and vulnerability. Put yourself in the learner's position and picture the scene: you are in a new environment (a business meeting, a seminar, a conference of 30 or 40 people); you need to ask questions and, aiming to engage rather than bore or frustrate your audience, you limit yourself to asking just two questions. Whatever the nature of the questions, they will in part be moderated by your:

- skills as a communicator, formulating clear, concise, well-framed questions that are focused on issues germane to the discussion and business of the meeting;
- critical verbal reasoning in shaping the question;
- ability to process information, so that the question acknowledges and goes beyond data already provided;
- capabilities with data, making informed judgments based on numerical evidence where this is appropriate;
- sense of empathy with the other people in the audience;
- recognition of the group's purposes, strengths, goals;
- professional awareness;
- self-awareness and personal confidence.

Assuming the conditions are right to ask a question in the first place, any hesitation in one of these abilities may lead to there being just the one question – the second may never get asked. To this extent, question-asking can be seen as a balance between cost and benefit.

To ask a question can mean incurring considerable costs in: (1) time; (2) mental and physical effort; and (3) confidence, esteem, status, embarrassment and peer judgement. This is not just in terms of asking the question, but also in terms of the consequences – having to decipher and understand the answer. There is a cost, too, if the answer is unfulfilling, unrewarding and lacking in usefulness.

It might come as no surprise, then, that students seldom ask questions in the classroom, commonly keeping their doubts and uncertainties to themselves (Dillon 1988; Maskill and Pedrosa-de-Jesus 1997; Watts and Pedrosa-de-Jesus 2011). There is a complex decision-making process at play – whether to ask a question or not: to balance achievement, personal development, enhanced status and weigh this against any increase in advantage. And, is there an acceptable 'social number' – an optimum number of questions – that might be asked in any one social context? That is, too few would be unfulfilling, and too many questions would be tedious, embarrassing and inappropriate. Not too few and not too many would provide a pleasing question-and-answer context. Or is it a balance between 'knowledge hunger' and 'satiation'?

To overcome some of these difficulties, Maskill and Pedrosa-de-Jesus (1997), Silva (2002), Pedrosa-de-Jesus *et al.* (2001), Teixeira-Dias *et al.* (2005) and Neri de Souza (2006) all suggest that teachers might use not learners' oral questions, but their written ones. These can be used as a 'secure' and private way of exposing doubts and knowledge gaps, providing also a longer time for reflection than the short time commonly prevailing in oral interactions (Maskill and Pedrosa-de-Jesus 1997). Etkina (2000) discussed the use of a weekly report, a structured journal in which students answered three questions: (1) 'What did you learn this week?'; (2) 'What questions remain unclear?'; and (3) 'If you were the teacher, what questions would you ask to find out whether the students understood the most important material of this week?'. Etkina's suggestion was that this not only encourages students to think about the gaps in their current knowledge, but also serves as an assessment tool, and allows the instructor to modify subsequent instruction to address students' needs.

There are two difficulties here. First, do teachers actually welcome students' questions? Rop's (2002) work suggests that teachers are ambivalent at best; they commonly listen to a student's open curiosity with contradictory feelings: the need to 'honour' a student's question, give students time and emotional support in their struggle to understand content, weighed against the seemingly incompatible need to maximise 'teaching time' and 'cover the content of the course'. Less compassionate, though, is when teachers perceive student questions as annoying, overbearing and a test of patience, particularly when they arrive too frequently or seem set to challenge received 'curricular' wisdom. In Rop's view, a 'question from the floor' can be facilitative or obstructive to a teacher, depending on whether it 'goes with the flow' of the lesson or 'goes against the grain'. The first enables the teacher to expand an argument or develop a topic, while the second is an irritation or impediment to a planned sequence and direction of the session.

Second, there is little in the discussion so far on how subsequent lessons were actually modified. Teixeira-Dias *et al.* (2005) do describe how they used students' questions as a springboard for instructional interventions. For example, they collected questions using a 'question box' in classrooms for anonymous written questions, and an email facility for 'out-of-hours' questions. In this instance, teachers were able to respond to direct queries, and created a series of special lecture sessions to tackle obstinate issues. Students broadly welcomed these, although the pace of curricular change was difficult to maintain. In other studies (e.g. Kulas 1995), students recorded their 'puzzle questions' in a diary or learning journal, setting out their 'I wonder' questions; Dixon (1996) used a 'question board' to display students' questions and suggested that these questions could be used as starting points for scientific investigations. In Watts *et al.* (1997), the authors argued for a question 'brainstorm' at the start of a topic, a 'question box' on a side table where students could put their questions, turn-taking questioning around the class where each student or group of students must prepare a question to be asked of others, and 'question-making' homework.

Two main gaps in this literature, then, relate to (1) the provision of specific in-lesson time for students' question, and (2) a systematic approach to using student questions as part of the systematic plans of forthcoming lessons. In Watts *et al.* (1997), the authors suggested including specific times for questions, such as a period of 'free question time' within a lesson or block of lessons, but failed to carry this through and explore the outcomes. With Teixeira-Dias *et al.* (2005) we designed instructional interventions, but lacked a systematic form of iterative lesson planning. In the study described here, we sought to remedy both of these deficiencies by testing the situations that best generated questions with our students, providing vehicles for question-asking in class, and then acting thoughtfully and reflectively on the questions they asked.

A classroom study

The research we discuss here was conducted with two parallel classes and a total of 54 pupils (Year 7, 12–13 year-olds) in a very normal secondary school. The timescale of this work covers 14 one-hour lessons across a four-week period of high school chemistry. Both classes were taught by the same chemistry teacher (Leite) and the curricular content addressed was 'physical and chemical transformations', 'physical properties of materials' and the 'separation of mixtures' – all fairly traditional components of the science curriculum at this level. Within the study, we identify three distinct phases: (1) an initial exploratory phase; (2) a question-generating phase; and (3) an implementation phase. At all points, students' written questions were collected and their oral questions were audio-taped within classroom lessons.

The exploratory phase

The key observations of both classes during the exploratory phase related directly to problems of pupil non-participation – very few students asked very few questions. There are three aspects to this:

- Pupils' participation was low in general, and this was reflected in the very sparse level of their question-asking.
- Most student contributions followed a request from the teacher; few of them were spontaneous.
- Those spontaneous contributions that did arise almost alwayscame from the same pupils.

The question-asking phase

In this second phase, we made efforts to increase pupils' interaction. Specific actions were designed and incorporated in lessons to encourage pupils to be more involved, to participate, ask questions and explore their ideas. These actions related to: (1) students' oral questions; (2) written questions;

(3) 'Question Moments'; (4) group discussion time; and (5) oral and written questions.

1 Students' oral questions

These were collected by audio-recording and transcribing the process of the classes.

2 Written questions

Three changes were made to the usual conduct of chemistry lessons. These were:

- *Question Sheets* – These were introduced at the start and collected at the end of every lesson. All questions were then read by the teacher and analysed for content and issues.
- *Group open-ended questions* – We created a set of written open-ended questions, handed out by the teacher to groups as group work, designed to foster discussion and hypothesising. Open-ended questions are defined as questions that have multiple possible answers. These kinds of question allowed students to take their previous experiences into their explanations, and also require students to justify their statements and explain their underlying logic (Lund and Kirk 2010).
- *Closed questions* – We also created these in order to identify possible gaps at the conceptual level. The written answers given by students during the group work were also gathered and organised according to the key ideas they presented.

3 Question Moments

For us, a 'Question Moment' is a pause in proceedings to allow students to write their questions and doubts. While Question Moments were intended as short breaks in normal class time to stimulate students' individual questions, students were nevertheless encouraged to write and ask questions whenever they wished. Importantly, they could also write questions about any topic – even if that had nothing to do with the matters addressed in the classroom. We expected these questions would provide some insight into pupils' wondering, reasoning, doubts and difficulties.

During this second phase we wanted to explore the effect of such Question Moments and so varied the inclusion of Question Moments. As Table 9.1 shows, Class A had two lessons where Question Moments were provided and Class B had one. The results are clear: perhaps unsurprisingly, the number of written questions is substantially higher in the lessons where specific Question Moments were provided, far outstripping lessons where no moments were provided.

Table 9.1 Number of written questions per lesson from two classes

	Class A lessons				Class B lessons		
	16 March	30 March	6 April	27 April	22 March	29 March	26 April
Question Moments	Yes	Yes			Yes		
Written questions	37	7	1	1	49	1	8

4 Group discussion time

After working in groups, pupils shared their answers to the open-ended questions with the class, presenting their arguments for and against each of the alternatives presented. The main aim of this discussion was to choose procedures that could provide solutions for the problems addressed by the open-ended questions. We hoped these discussions would also provide greater insight into pupils' minds, as well as allowing them to put their ideas to the test by confronting them with those of their peers.

5 Oral and written questions

Each class was divided in two (Shift 1 and Shift 2), each having lessons at different times of the school day.

In both lessons, the number of pupils who wrote questions was higher than those who made oral interventions. This means that some of the pupils who did not participate orally during those lessons wrote questions, suggesting that those pupils were more disposed to write rather than to voice their doubts out loud.

The implementation phase

In this phase we adopted two main approaches. First, all the oral and written questions, and the group open and closed questions, were used to gain insight

Table 9.2 Pupils' oral and written interventions

	Class A, 16 March (14 pupils)		Class B, 22 March (13 pupils)			
	Shift 1 interventions		Shift 1 interventions		Shift 2 interventions	
	Oral	Written	Oral	Written	Oral	Written
Number of interventions	77	25	156	28	136	21
Pupils who did not intervene	5	1	3	0	4	2

into pupils' states of knowledge, their doubts and possible lack of knowledge and understanding. The problems identified in this way then served as a basis for planning the next lessons. Second, these questions were presented to the class as matters for discussion, helping to revisit topics previously addressed, and to clarify doubts.

We hoped that integrating pupils' questions (and answers) of this kind into the lesson would increase their motivation to participate. Given the number of questions generated, it became necessary to set criteria for selecting the ones that would be discussed. Our choice was supported principally by two criteria: the number of pupils who expressed similar doubts, concerns or curiosities (the most frequently asked questions); and the relevance of those worries to the organisation of, and the approach to, the curriculum matters in hand.

A first example of this is the lesson of 30 March when, from a total of seven written questions, four were focused on the phenomenon of sublimation. These questions were:

- How can it [a substance] pass from the solid state into the gaseous state and vice-versa?
- How can the solid state transform into gaseous and vice-versa, and not pass through the liquid state?
- How can it [a substance] pass directly to the gaseous state?
- How can ice transform into water vapour and not pass through the liquid state, and vice-versa?

These are essentially the same question, and about a relevant aspect of the topic – physical transformation. Therefore we thought it was important to promote class discussion on this in order to address these questions.

A second example is drawn from the lesson of 22 March, when two similar written questions were received about the concept of pressure:

- I did not understand the explanation about pressures.
- I did not understand very well that part of the lesson about pressure.

These two students had simply not understood the explanation offered to them. There was no oral intervention (no question or comment of any kind) from any student during either class; no questions asked of the presentation of the concept of pressure at the time. One possibility is that the concept was so abstract that these two students (at least) felt too uneasy even to comment or question it at the time. So, although these two questions were only a small fraction of the whole set of questions (49), we thought it was appropriate to discuss this concept again. Besides, pressure is a term used to define boiling and fusion points – both central concepts of the programme – and therefore we thought students needed a 'palpable' understanding. This, then, promoted a class discussion on the concept, using the example of a chemical reaction, where a balloon was inflated because of the pressure exerted by a gas formed during the reaction.

Outcomes

The initial exploratory phase allowed us to identify two groups of pupils whose preferences were distinct: the minority of 'oral askers' who would speak out and ask questions in class, air their doubts and reasoning; and the majority of 'question writers' in the two classes, who preferred writing their ideas in a more individual and private manner. These observations do support the idea put forward by several authors (Dillon 1988; Maskill and Pedrosa-de-Jesus 1997; Watts and Pedrosa-de-Jesus 2011) that pupils *do* have questions to ask and are able to ask them if the right conditions are provided.

Table 9.1 above shows that the inclusion of a specific Question Moments for writing questions in class clearly favours the writing of questions. Interestingly enough, even in the lessons where no such moments where provided, some students did write questions, which suggests that the Question Sheet alone was also a meaningful tool for these students. The moments for writing produced a varied set of questions (a total of 105 over the course of the study). Most of the questions were acquisitive of factual information, and concerned concepts or terms used in the classroom. Some examples:

- How can we know if the transformation is chemical or physical?
- What is the name of the temperature symbol θ?
- What is a reagent?

There were a few questions, however, which were wider, and tried, for example, to establish connections between issues learnt in the classroom and pupils' previous knowledge, or imagining scenarios as a way of 'testing' the new information:

- When water passes into the gaseous state, does it disperse in the atmosphere? If so, can we say that an ice cube is created from the liquid water that came from the gaseous state of several places?
- If water modifies its physical state (when it is too hot it evaporates and when it is too cold it solidifies), why are there more clouds in winter, when it is colder?

The number of questions written during two of the lessons where Question Moments were provided was compared with the pupils' oral interactions during the same lessons. Table 9.2 above shows the results.

As noted earlier, in the lesson on 30 March, four of the seven written questions were related to the idea of sublimation, the topic taught for the first time in that lesson. In that case, to help pupils understand the phenomenon, the teacher decided to break with her schedule in the next lesson to demonstrate sublimation with a sphere of naphthalene, since this sublimates at room temperatures and was a familiar example for pupils. The classroom talk below contains a small extract of the subsequent discussion. This exchange, built upon the pupils' initial written questions, addressed their doubts and also allowed other pupils to express their ideas and reveal misconceptions. For example, the idea of 'the physical transformation of the naphthalene from solid state into gaseous state' as a *transmission of particles* was deconstructed and given adequate feedback for a better understanding of the phenomenon.

T: [...] Today I brought you a substance that everybody knows and which sublimates easily. [...] Have you never seen these little balls in your wardrobes at home?

S1:   Aaaah, I have!

T: It is usually used to repel moths ... and other insects and it is called naphthalene. [...] Well, what evidence do we have that naphthalene is passing from a solid into the gaseous state?

S2:   It is slowly disappearing.

S3:   What?

T: Well, the naphthalene ball will get smaller and smaller, yes, and what else? There is another piece of evidence.

S3:   [...]

T: The smell! If I leave this ball here for a while, if I leave it here in this corner, in a while you students will be able to detect the smell of naphthalene back there, in the other side of the room. Why is that? What do you think that happens?

S1:   Because while ... the ... the naphthalene is transmitting ... its particles.

S2:   Because it will pass into the air.

S4:   [...]

T: Yes, because sublimation has occurred!

Some of the naphthalene molecules have passed from the solid state into the gaseous state, haven't they? Be careful, it didn't occur, how did you say Pedro, a ... a transmission of particles, but a change in the physical state of some of the particles of naphthalene – from the solid state into the gaseous state. In which of the physical states will the molecules of naphthalene have a greater mobility – in the solid state or in the gaseous state?

S1:   In the gaseous state.

S5:   Solid.

S6:   No, gaseous!

T: Exactly! They will have a greater mobility in the gaseous state and can go from here to Maria's nose!

All: [Laughter]

T: This is evidence that there was a physical state change. But at no moment do we see naphthalene in the liquid state, right?

S7:   No, we don't.

T: It is for that reason that we can use naphthalene to keep moths away from our clothes – because it is never in the liquid state, it does not wet our clothes.

Below is an example of another discussion prompted by students' written questions, this time about gaseous pressure. Because pressure is a term used to define boiling and fusion points, we thought it necessary to give students a 'palpable' understanding of what this means. So, we used a question written by one student to initiate discussion, and also to remind students of what they had observed in the lesson 22 March, in the thermolysis reaction of sodium hydrogen carbonate.

Student's written question (22 March):

- In the balloon experiment, if we allowed it more time, would it burst?

Class discussion on students' written questions (29 March):

T: Do you remember the chemical reaction we saw in the last lesson, which took place through the action of heat?

S8:   It was the one with the yeast!

T: Ah, so you still remember! Yes, that's the one, the reaction of decomposition of sodium hydrogen carbonate, which is a constituent of yeast. And what were the products of that reaction?

S9:   Water.

T: Right, and what else?

S10:   Carbon dioxide gas.

T: Very good! And how did you … what proof did we have that that gas was forming?

S11:   The balloon swelled.

T: Right, that's the reason why we put a balloon in the flask, wasn't it? To detect the formation of gas. Now, one of you wrote this question: 'In the balloon experiment, if we allowed it more time, would it blow up?' What do you think?

S10:   I think it would blow up.

S12:   It would need a lot of gas.

T: It could eventually blow up if the pressure that the gas particles exerted over the balloon wall was too strong. Do you remember what we said pressure was?

S: […]

T: We saw it as a force exerted on a certain area. In this case it is the force … exerted by the gas particles when they bump against … imagine … a square centimetre of the balloon rubber. The bigger the number of particles, the bigger ... the bigger is the force per square centimetre, that is, the bigger the pressure.

S13: And the bigger the pressure, the more it stretches! Is that right?

T: Exactly! The pressure may be so big that the rubber of the balloon can't stretch anymore and the balloon blows up.

S10: I said it would blow up.

The first lesson about pressure elicited nothing but silence. In this second one, the students felt more comfortable to comment while trying to make sense of what was being said ('I think it would blow up', 'It would be needed a lot of gas', 'And the bigger the pressure the more it stretches! Is that right?'). This questioning approach made the topic more accessible to students.

Within this strategy, it is worth noting that many of the questions that students wrote concerned matters discussed in lessons that had taken place two weeks before. This means that those students had kept their doubts to themselves for all that time and exposed them only when they had the opportunity to write them down.

Open-ended and closed question Activity Sheets

The group Activity Sheets used in the lessons comprised a total of six open-ended and three closed questions. Four of the open-ended questions could be answered in (at least) two different, yet valid ways. Our approach was that this activity was best suited to group work, so that there were many heads developing an answer to the question. They could also explore the issues involved by raising their own questions. Some of the proposals that students made – and that could be considered viable – had not been foreseen by the teacher. For instance, two groups suggested separating a mixture of flour and iron filings by adding water to the mixture:

- We could separate the iron filings from flour using water because the flour would stay at the surface and the iron filings would sink.
- To separate the flour from the iron, we would have to put them in glassware with water. One stays on top and the other stays at the bottom.

That is, students thought of taking advantage of the different densities of flour and iron filings. Although the density concept had not been formally taught, students seemed to have an intuitive understanding of the different behaviour that the two components of the mixture would display in the presence of water. Alluding to this proposal in the next lesson was then a useful 'starting point' for the introduction of the concept of density, incorporating students' own ideas as part of the lesson plan. The 'expected' answer in this instance was the use of a magnet, and it was, indeed, the answer given by the majority of students (12 groups):

- We could use a magnet and it would attract the iron filings.

Class discussion was then used to help choose the more efficient means of sepa-rating, the components of the flour and iron filings mixture. This process increased participation and involved a greater number of students, because it incorporated their different ideas and presented reasons for choosing one pro-posal over the other, instead of dismissing or simply neglecting the 'unexpected' ideas. Below we present an extract of one of those discussions.

T: So, to separate this mixture, what material will be needed?
S14:   Water!
S15:   A magnet!
T: Ok, we have here two different ideas. Why do you say water, S14?
S14:   I think it's because …
S16:   Because if we put water [on the mixture], the flour floats.
T: Anyone else thinks as this group?
         [...]
S17:   It's with a magnet!
S18:   Miss, we think that with a magnet we could attract the iron filings because
         … since iron filings have magnetic properties and the flour doesn't.
S16:   Ah, maybe we are wrong …
S19:   No, I think the iron sinks and the flour stays on the water. It stays like in
         this one [mixture of water, flour and sand].
T: Yes, in both cases, the mixture of water, sand and flour and this mix-
         ture [sugar and copper sulphate], if we add water, the components
         denser than water will stay in the bottom. In this case it is the sand and
         in this case it would be the iron filings. But look, notice that the flour
         in this mixture [water, sand and flour] does not stay all floating in the
         water, some of it absorbs so much water that it eventually deposits at
         the bottom. You saw that some of the flour deposited in that mixture
         [mixture of water, sand and flour] and that's why we had to stir it,
         wasn't it? So, we would have here a problem. As for the other sugges-
         tion, the magnet …
S19:   With the magnet we attract.
T: Hmm, explain a little further S19.
S18:   Filings have magnetic properties.
T: What has magnetic properties?
S19:   The iron.
T: The iron filings have magnetic properties. And the flour?
S20:   The flour doesn't have, that's why we can separate: one has and the other
         doesn't.
T: That's right. Because iron fillings have magnetic properties and flour
         doesn't, we can separate these components using a magnet. This is a more
         practical way of separating the mixture than the other one, suggested by
         that group, since we don't have the problem of flour depositing over the
         iron fillings, as it would happen if we add water [to the mixture].

A further question on the group work sheets asked students to explain why salt is spread over snow in the streets in winter, and then to present a way of testing their explanation. We were ready to welcome all the meanderings and doubts this question might create and knew they would probably be a good matter for discussion. Below we show an excerpt of class discussion.

T:	One of the questions asked you to suggest an experiment that allowed you to test if your explanation was valid. All the groups suggested putting ice in a glass beaker, put some salt on top of the ice and wait to see if the ice would change from the solid state to the liquid state. Correct?
Class:	Yes.
T:	What did you expect to happen?
S21:	The ice will melt.
T:	Ok, if we put some ice in a beaker here in the classroom and wait, what do you expect to happen after a while?
S22:	The ice will melt.
T:	Ok, it would change from the solid state to the liquid state. So, do you think your suggestion helps you understand the effect of pouring salt over the ice?
S23:	Nope!
S22:	It is the same.
S23:	There's no difference.
S24:	It has to do with the temperature. The problem is the temperature! We explained it has to do with the temperature!
T:	Right! The important difference between these two cases is the temperature of the environment.
S21:	If it was cold, there would be differences between the ice with salt and the ice alone.
T:	Ok, what differences would those be?
S21:	I mean, if we were at $-5\,^{\circ}\mathrm{C}$ and we had just ice and no salt, then ice wouldn't go from the solid state to the liquid state. Because ice alone just changes to the liquid state at $0\,^{\circ}\mathrm{C}$.

These examples show how these strategies stimulated and increased pupils' participation, increasing pupil-teacher classroom cooperation, and helping the teacher to individualise teaching, that is, to be aligned with pupils' learning preferences.

The analysis of the questions allowed us to characterise their main misunderstandings of the concepts involved, choosing some we considered relevant to stimulate the discussion in the following lesson. For example, on 22 March, a lesson dealing with the concepts of boiling and fusion point was planned, taking into account the answers already provided for the following question:

- Why is water in a solid state at normal pressure and ambient temperature of −5°C?

The answers required pupils to understand the concept of freezing point. So, we decided to begin the discussion using one of the answers to that question:

- Water is in the solid state *because* the room temperature is −5°C.

The discussion below is a small extract from that raised by that answer.

T: [...] One of the groups wrote this answer: 'Water is in the solid state because the room temperature is −5°C.' So when the room temperature is −4°C, is the water no longer in a solid state?

S25:   No, it is. That happens only if the temperatures are negative.

T: That is, at any temperature below 0°C?

S26:   Yes, it has to be lower than the fusion point of the water.

T: Ah! That was the point most of the answers lacked! Let's see, imagine that here in the classroom we are at 15°C. Why is sodium chloride here in a solid state, and yet the water is in a liquid state? Look carefully at the list.

S26:   Because the temperature needed to ... melt ... to fuse sodium chloride is 801°C.

T: Very good! Now identify a substance that would be here in the room in a gaseous state.

S27:   Oxygen.

T: And your justification for that is ... ?

S27:   Because the boiling point of oxygen is lower than the temperature that ... that we have here in the classroom.

The rolling programme

The classroom discussions above are examples of a 'rolling programme' of revisions and classroom interventions. In this instance, we modified the lesson on 30 March (Table 9.1) in the light of the previous lesson, then the lesson on 6 April in the light of 30 March; the lesson on 27 April after 6 April, and so on – a series of previously unplanned departures from our curriculum schedule of lessons as a result of students' questions. Our reflective log at the end of each lesson served as a basis for the next lessons. For instance, in one lesson, problems were detected concerning verbal expressions such as 'vanishing of substances', concerning both physical and chemical transformations. We used a question written by one of the students in that lesson to tackle this problem in the next lesson, and to reinforce the idea of conservation of mass during transformations.

Student's written question (16 March):

• The quantity of water remains the same when it goes from one state to another or the water disappears?

Class discussion on pupils' written questions (30 March):

T: Now, another question … There was a student who asked if the quantity of water changed when the water changed state or if the water disappeared. Who would like to answer this question?
      […]
S28: I think it is always the same.
T: You think it remains the same. Are there any more opinions?
S29: If we have a … a closed box with an ice cube inside and we increase the temperature, it melts and then passes to the gaseous state, but the quantity of water inside the box is still the same.
T: Exactly, the quantity of water remains the same. But what if the box is open?
S28: What?
T: What if the box is open?
S30: Hmm … the water will spread, won't it? But the quantity of water is still the same.
T: Exactly, the water in the gaseous state will leave the box, but the mass of water still inside the box plus the mass of water in the gaseous state that has left the box is equal to the initial mass of water that was originally inside the box.
S28: […]
T: Yes, Teresa?
S28: The quantity of water is still the same, but the density changes.
T: Very good! […] Could you explain to us what do you understand by density?
S28: For me, density is a mass in a certain volume. It can be more concentrated or more diluted – there can be more mass in the same volume or less, what means more or less density.
T: Very good Teresa! It is true that in these two examples, open box and closed box, when … at the moment we open the box, the water vapour that was inside can distribute into a greater volume and therefore the density of that water vapour decreases. Hmm … by the way, thinking about the water, which state do you think corresponds to a greater density – the liquid state or the gaseous state?
S28: The liquid state.
T: Correct! Because in the liquid state of the water, the same mass occupies a smaller volume.
S31: The mass is the same, the volumes is the one that changes.
T: Exactly! The mass is the same; the volume is the one that is different. 100g of water vapour occupies a greater volume than 100g of liquid water, doesn't it?
S29: Yes.

S30:   Yes.
S28:   In the air it is more spread.
S32:   That's what it means to be less dense.

In this discussion students revealed good reasoning on the topics being discussed, and it is worth nothing that density was not a topic that had been addressed before in the class.

In some of the lessons (for example 6 April, 26 and 27 April), the Activity Sheets were used, and the discussions in those lessons centred on students' oral inquiries and the written answers given there and then to the questions on the sheets. Lesson time was tight: to both undertake the activity and maintain momentum within the curriculum, in some instances, students began to answer the questions on the sheets in one lesson (22 March) but finished them in the next (29 March). Time was made available in the latter lesson to discuss student's written answers.

Conclusion

Reflection upon students' oral and written questions, and subsequent classroom discussions, has provided us as teacher-researchers with relevant information about individual and group 'knowledge gaps', doubts and perplexities, both implicit and explicit. The teacher created time in subsequent lessons outside of her normal curriculum planning to address problems identified, and to promote class discussions initiated by presenting the pupils' 'questioning products'. The exemplar data above illustrates how the teacher used the collected information and managed discussions, based upon pupils' written questions and written answers to trigger conversations.

There is no doubt that promoting learners' questioning demands more of the teacher than simply lecturing to a sea of blank faces – the price for prompting engagement and inquiry in the classroom is the development of a set of planning and organisational skills, alongside a willingness to receive, analyse, evaluate and address their questions. These examples are merely illustrative and serve only to illuminate some of the key issues involved in inquiry-based teaching practice. The data do not build theory but provide empirically-grounded contexts for demonstrating both the advent of classroom questions, descriptions of the situations that give rise to them, and some of the possible ways in which teachers can manage them.

Key questions

It seems very appropriate to have questions at the end of a chapter about questions. So:

1   Can you recall those classroom questions that have 'stumped' you and you have had to buy time in order to get an answer, if only for yourself? In the literature, classroom moments like that are called 'critical incidents' and there is a series of papers about them (see below).

2   How do you deal best with questions? Can you sort the 'chaff' from the 'grain' – those questions that some youngsters will use to set up 'red herrings' in the lessons from those that are genuine and really need a good clear answer?

Further reading

Watts, D. M., and Pedrosa-de-Jesus (2011) Questions and science. In R. Toplis, *How science works: Effective pedagogies and practice for secondary school teachers.* London: Routledge. This is a good introduction to some of the issues we have been discussing here.

Pedrosa-de-Jesus, M. H., and Watts, D. M. (2012) Managing the affect of learners' questions in undergraduate chemistry. *Studies in Higher Education*, 1–15, iFirst Article. This paper takes the issues further and develops a theoretical background to the asking of questions.

Browne, N. M., and Keeley, S. M (1998) *Asking the right questions: A guide to critical thinking.* New Jersey: Prentice Hall. An old but excellent and easy-to-read text that moves the discussion into critical thinking.

References

AAAS (American Association for the Advancement of Science) (1989) *Science for all Americans.* New York: Oxford University Press.

AAAS (American Association for the Advancement of Science) (1993) *Benchmarks for science literacy.* Available online at: http://www.project2061.org/ (accessed 4 November 2013).

Brush, T., and Saye, J. (2000) Design, implementation, and evaluation of student-centered learning: A case study. *Educational Technology Research and Development,* 48(3): 79–100.

Chin, C., Brown, D. E., and Bruce, B. C. (2002) Student generated questions: A meaningful aspect of learning in science. *International Journal of Science Education,* 24(5): 521–49.

Cuccio-Schirripa, S., and Steiner, H. E. (2000) Enhancement and analysis of science question level for middle school students. *Journal of Research in Science Teaching,* 37: 210–24.

Dabrowska E., and Lieven, E. V. (2005) Towards A lexically specific grammar of children's question constructions. *Cognitive Linguistics,* 16: 437–74. doi:10.1515/cogl.2005.16.3.437

Dillon, J. T. (1988) *Questioning and teaching: A manual of practice.* Beckenham: Croom Helm (pp. 6–41).

Dixon, N. (1996) Developing children's questioning skills through the use of a question board. *Primary Science Review,* 44: 8–10.

Dori, Y. J., and O. Herscovitz, O. (1999) Question posing capability as an alternative evaluation method: Analysis of an environmental case study. *Journal of Research in Science Teaching,* 36(4): 411–30.

Etkina, E. (2000) Weekly reports: A two-way feedback tool. *Science Education,* 84(5): 594–605.

Good, T. L., Slavings, R. L., Harel, K. H., and Emerson, H. (1987) Student passivity: A study of question asking in K-12 classrooms. *Sociology of Education*, 60: 181–99.

Harper, K., Etkina, E., and Yuh-Fen Lin (2003) Encouraging and analyzing student questions in a large physics course: meaningful patterns for instructors. *Journal of Research in Science Teaching*, 40 (8): 776–91.

Kulas, L. L. (1995) I wonder .... *Science and Children*, 32(4): 16–18.

Lund, J. L., and Kirk, M. F. (2010) Open-response questions. In S. Quinn (ed.) *Performance-based assessment for middle and high school physical education*. Windsor: Human Kinetics (pp. 95–110).

Maskill, R., and Pedrosa-de-Jesus, H. (1997) Pupils' questions, alternative frameworks and the design of science teaching. *International Journal of Science Education*, 19(7): 781–99.

Neri de Souza, F. (2006) *Perguntas na aprendizagem de Química no Ensino Superior.* Tese de Doutoramento não publicada. Aveiro: Universidade de Aveiro.

Paul, R., and Elder, L. (2007) *Critical thinking: Concepts and tools.* Tomales, CA: Foundation for Critical Thinking.

Pedrosa-de-Jesus, H., and Watts, D. M. (2012) Managing the affect of learners' questions in undergraduate chemistry. *Studies in Higher Education*: 1–15. iFirst Article.

Pedrosa-de-Jesus, H., Neri de Souza, F., Teixeira-Dias, J. J. C., and Watts, D. M. (2001) *Questioning in chemistry at the university.* Paper presented at the 6th European Conference on Research in Chemical Education, University of Aveiro, Aveiro, 4–8 September.

Rop, C. J. (2002) The meaning of student inquiry questions: A teacher's beliefs and responses. *International Journal of Science Education*, 24(7): 717–36.

Silva, M. R. P. (2002) *O desenvolvimento de competências de comunicação e a formação inicial de professores de Ciências: o caso particular das perguntas na sala de aula.* Dissertação de Mestrado não publicada. Aveiro: Universidade de Aveiro.

Teixeira-Dias, J. J. C., Pedrosa-de-Jesus, M. H., Neri de Souza, F., and Watts, D. M. (2005) Teaching for quality learning in chemistry. *International Journal of Science Education*, 27(9): 1123–37.

van der Meij, H. (1994) Student questioning: A componential analysis. *Learning and Individual Differences*, 6(2): 137–61.

Watts, D. M., and Pedrosa-de-Jesus, H. (2005) The cause and affect of asking questions: Reflective case studies from undergraduate sciences. *Canadian Journal of Science, Mathematics and Technology Education*, 5(4): 437–52.

Watts, D. M. and Pedrosa-de-Jesus, H. (2011) Questions and science. In R. Toplis (ed.) *How science works: Exploring effective pedagogy and practice.* London: Routledge (pp. 85–102).

Watts, D. M., Alsop, S., Gould, G., and Walsh, A. (1997) Prompting teachers' constructive reflection: Pupils' questions as critical incidents. *International Journal of Science Education*, 19(9): 1025–37.

Chapter 10

Explanations and explaining in science: you can't force people to understand

Mike Watts

> We live in a society absolutely dependent on science and technology, and yet have cleverly arranged things so that almost no one understands science and technology. That's a clear prescription for disaster.
>
> (Carl Sagan quoted in Heald 2006)

Introduction: what is understanding?

'Sir?'

'Yes, Sal?'

'Why is the Earth round?'

This chapter explores what it means for a teacher to explain something in science, explanations that are targeted at understanding, what is involved in the act of explaining, and why explanations and understanding are important. And why all this is controversial and very much a matter of debate.

Typically, scientific explanations are answers to 'why' questions such as Sal's question above about the earth. Explanations relate to the position in which we find ourselves when we know something but not everything, and some of the things we do know about we do not understand. And, in order to understand that thing, we need a satisfactory explanation. Science is a subject without end, and there are many and varied events, objects, rules, concepts, facts, principles, laws in science that require explanation – no matter what stage of learning and understanding we have reached. Alongside that, there is an enormous literature on the nature of explanation in the philosophy, the psychology and the practices of science. In this chapter I try to wade through some of the explanatory quagmire.

To begin: there are many different types of explanation and, for example, back in early Grecian times Aristotle recognised at least four basic types. There have been many more types since his pronouncements and I consider just a few of them here. The direction I follow in this chapter is that explanation can best be understood in the general context of inter-personal communication: it is people who construct explanations – facts alone do not. A fact simply cannot, by itself, explain something. So, for instance, for a teacher to answer a question with the single word 'osmosis', 'distillation' or 'gravity', or an expression like 'Not round but oblate', is not to give any explanation at all. 'You seem very clever at explaining words, Sir', said Alice. 'Would you kindly tell me the meaning of the poem *Jabberwocky*?'

So, for my purposes here, an explanation is a response to a question posed within a particular social context (for example, a school science classroom, the school grounds, a field study centre, a science museum) by somebody asking for and about certain ideas or information. For a longer debate about the nature of context, have a look at John Gilbert's discussion in Chapter 11. An explanation would, if it happens properly, fulfil certain of the questioner's intellectual goals. At their very best, scientific explanations 'light up our minds with a bolt of insight' (Strevens 2009, 136). To understand a phenomenon is to have it explained: explanations generate understanding, they help the question-asker to 'make sense' of something, satisfy their need to understand. In this way, explanation has its root in the practice of raising questions and giving answers, with the expectation of being answered genuinely in one way or the other so that it gives us a strong psychological feeling of knowing, of having increased our grasp of the matter in question, of having constructed some sensible meaning for what is going on. If we already know something, then we don't need an explanation: if someone then sees fit to give us an explanation, it adds nothing new to what we already know. If I, as the questioner, do not experience that I learn something new, then the explainer has not, in my eyes, actually provided an explanation. So, what counts as an explanation for one person is not necessarily an explanation for another. Only in the situation where the explanation 'fits' into the question-asker's background knowledge, his or her schema, or 'question frame' will it provide an insight and a newly constructed meaning. On this basis, we teachers are commonly in the position of giving explanations that are not explanations at all. Table 10.1 shows a range of responses adapted from those suggested by Webb *et al.* (1995) in relation to problem-solving in science.

Table 10.1 Definition of explanation

Term	Definition
Explanation	A verbal description that articulates the major principles necessary to solve a problem
Minimal explanation	A description of only some of the major principles necessary to solve a problem, or a reference to a resource that will point a peer in a direction but does not itself describe the underlying principles or conceptual structure
No explanation	A response that answers the question without elaborating on a reason why the answer is correct

Source: Webb *et al.* (1995).

The act of explaining is natural, just as it is spontaneous. Young children give explanations for all manner of phenomena as a matter of course and the early work of Jean Piaget, the Swiss epigenesist charted his delight with the explanatory frameworks children developed in response to his questions (Piaget 1929). Some of their explanations are what psychologist Tania Lombrozo (2010) calls 'promiscuously teleological': explaining things by the purpose they serve instead of digging deeper for meaning (i.e., young children are more likely to say that a mountain exists to be climbed and not because of some geological forces that happened to shape the earth a certain way). And, argues Lombrozo, we never really outgrow this childhood tendency – in fact, we revert to it if we simply feel stressed or distracted or, more seriously, if we suffer cognitive decline through diseases such as Alzheimer's. When in doubt, our brain seems to take the easiest route to determining causality, and it does so quickly and authoritatively.

To 'explain explanation' also takes us to a discussion of 'process' and 'product'. An explanation is both: it can signify the act of explaining – or the result of that action. A simple example of this process/product duality is destruction: 'I saw the destruction at Croydon'. The sentence might mean either that I saw the act of Croydon being destroyed or that I saw the results of such an act. Most of the classical literature on explanation, certainly by the Greek philosophers Plato and Aristotle, deals only in the idea of an explanation as a product. They believed that an explanatory product could be characterised solely in terms of the kind of information it conveys, with no necessary reference to the act of explaining being required. My discussion here covers both the process and product because I am interested in the 'construction of meaning' by teachers and learners and, while the word 'construction' (like 'destruction') has both a process and product meaning, the act of teaching implies enabling learners to construct their own understandings of (in this case) parts of science. The components of an explanation can be implicit, and be interwoven with one another and, as always, are subject to interpretation and discussion. In science, explanation is one of the purposes of scientific research – a way to uncover new knowledge and to report relationships among different aspects of the things that are happening.

What are explanations in school science?

'Miss?'

'Yes, Chris?'

'My mother drinks lemon tea. She says that when she adds a few drops of lemon to her black tea, then the tea stops being black and changes colour to a light brown. Why is that?'

As teachers of science, we are confronted in our daily lives with a diverse range of learners. In a secondary classroom there may be ten (A-level), twenty (GCSE) or thirty (Key Stage 3) learners at any one time. Or more in each case, depending on school circumstances. And all learners are different: they bring different experiences, background knowledge, attitudes and motivation to the classroom. Thankfully, there are also many commonalities, and so teachers can, in general, expect certain kinds of background and approach from many of the students before them. The act of understanding, though, is individual and diverse. A teacher cannot force a learner to understand something; that learner alone must make the necessary internal, intellectual connections to 'switch on the light of understanding' for themselves. The connections they make must come from their own mental efforts. One role of the teacher is to support this process, to foster and enable each student's under-standing. He or she will do this by trying to explain.

Very often in science classes, teachers are asked to explain reasons and rationales of certain events. To that extent, an explanation is a set of statements constructed to clarify causes, context and or consequences. As the philosopher David Hume observed in 1740, 'Causality is the cement of the universe'. Explanations help to establish rules or laws, and may serve to clarify the existing rules and laws in relation to particular phenomena or other parts of science, try to show *why* or *how* something is or will be. So, for example, the teacher might respond to the student's question about their mother's tea, and explain that the citric acid of the lemon would react with the colourants in the tea, much as it would with other vegetable dyes: the veg-etable dyes in tea are weak acids, changing colour as the acidity changes. This is a rule of thumb, not all vegetable dyes will change colour.

The trouble with a why-question is that it is commonly ambiguous: the stu-dent might be asking, 'Teacher, why do you (or why should we) believe that world is round?' In this case the explanation comes in the form of evidence (the curve of the horizon, the shadow of the Earth on the moon's surface, pictures from space). These are the consequences of the Earth being round. But the question might also be 'Why is the Earth *round*?' That is, not square, oblong or irregular. And that, as they say, is a whole new ball game.

This latter point, the importance of understanding may, or may not, be self-evident. Of *course* science teachers work to explain phenomena in science – surely that's the very kernel of the job? Or is it? Countless young people emerge from school and have quite evidently failed to understand the explanations they have been given. As an example, Figure 10.1 below illustrates some representative common ideas about electric circuits. In Circuit 1, a battery is connected to two identical light bulbs A and B in series. In Circuit 2, the battery is connected to a single bulb C that is identical to bulbs A and B.

Back in 1992 McDermott and Shaffer (1992) asked physics students to compare the brightness of bulbs A and B in Circuit 1 and to compare these with the bright-ness of bulb C in Circuit 2. The results of this investigation have been replicated many times, in many contexts and many countries. The correct answer, that the two bulbs in Circuit 1 are equally bright and that the bulbs in Circuit 2 are equal to each other and brighter than those in Circuit 1, was given by only about 10 per cent of 16 year-olds in mathematics courses and by only about 15 per cent of the students

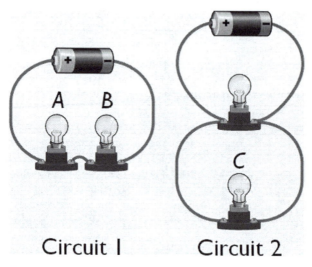

Figure 10.1 Bulbs in series and in parallel.

in science-based courses. The most remarkable result of McDermott and Shaffer's study was that the types of errors made by the pupils on this question are *unrelated* to, and *unaffected* by, conventional instruction. One common student error is the belief that in Circuit 1, one bulb will be brighter than the other because the first bulb in the circuit 'uses up' the current first. Another common error is that the brightness of each bulb will be the same in both circuits because the battery provides a constant *current* in all cases. Neither of these (incorrect) ideas is learned from a science course, but neither are they *discredited* in school science. Indeed, McDermott and Shaffer (1992) found that student performance on this question was nearly independent of whether the question was posed before or after instruction on electric circuits. Similarly disquieting results have been found regarding other ideas in many other parts of science. The best that some teachers can hope for is that their learners simply memorise enough to pass examinations. If they are sufficiently keen or needy to continue to study science beyond this, then, perhaps, the explanations will turn up eventually, will arrive and make sense later. Understanding will dawn in due course.

What do explanation and understanding have to do with emotion?

'Sir?'

'Yes, Pat?'

'Why do people have eyebrows?'

As Newton (2000: 6) points out, understanding satisfies a number of personal needs:

We often feel dissatisfied with merely knowing that event B followed event A. What we are curious about is *why* B followed A.

The suggestion here is that puzzlement, curiosity, perplexity, doubt, challenge, wonder, incongruity are all forms of what Piaget (1971) called *disequilibrium*. His basis was that human beings feel satisfied with a good match (an 'equilibrium') between our mental ideas, personal schemas, individual theories and the physical and social worlds we encounter. While perfection is approximate in most cases, and may never actually be attained, people in a state of equilibrium are generally seen to be *satisfied* with their modes of thought. However, we are frequently faced with disturbances: new events or situations, contrary evidence that cannot be fully handled by our existing understandings. This creates an imbalance between what is understood and what is encountered. Awareness of shortcomings in our existing thinking produces dissatisfaction so that *dis-equilibration* ensues, an uncomfortable state sometimes referred to as cognitive conflict. People then attempt to reduce such imbalances by focusing on the issues that cause the disequilibrium, and look for new explanations, develop new thoughts, ideas and understandings, or they adapt old ones, until equilibrium is restored once more.

For example, students are invited to imagine themselves on a long stretch of beach, facing the sea, watching the eastern horizon. The moment they are waiting for, watching, is the approach of dawn. Asked to say what they see over the next short while, they describe the Sun as rising in the sky above the horizon, a process that continues as it gains height through the day until, past noon, it begins its descent, and the Sun sets in the west. The teacher intervenes and contradicts their description. That's not what is in fact happening, no, not at all. What is actually taking place is that the Earth is 'dropping away', rotating down past the Sun: the Earth is moving relative to the Sun. In this case it is the Earth's movement that counts, not the Sun's. A better descriptor would be 'Earth-fall' rather than sunrise, and 'Earth-rise' rather than sunset. Coming to terms with this shift in perspective is called the processes of *accommodation* and *assimilation*: we adopt more sophisticated modes of thought that serve to eliminate the shortcomings of the old ones and, having done so, we are then considered to have found an explanation that satisfies our previous lack of understanding.

It is clear that feelings of satisfaction, dissatisfaction, comfort and/or discomfort about thoughts or explanations are emotional, affective issues, and not just cognitive ones. So dissatisfaction, or discomfort, with new incoming information is an emotional state. When a student sits in a lesson and is presented with small, medium or large variations to what they already know and hold meaningful, they will experience a range of small, medium or large emotions. These can be positive, negative or some combination of both. So, for example, small feelings of uncertainty, reservation, perplexity and apprehension might creep in as the lesson is being presented, even though these could also be accompanied by feelings of some appreciation of how the new ideas might fit together for the teacher – even if they feel mildly alien to the student.

There is no doubt, too, that explanations can arouse quite strong emotions – 'hot' emotions such as frustration, fear, revulsion, pleasure, hope and joy. It is not news to point out that people can object very strongly to what they perceive as unpalatable, insensitive, drastic or offensive ideas that are radically at odds with their own. Explanations that move understanding from the familiar to the unfamiliar have the potential to generate a vast range of emotional reactions (Cartney and Rouse 2006). As Claxton points out:

> Learning is generally a risky business because it means moving out from the safety of the known into the unknown and the controlled. [...] The involvement of emotion in learning, especially any that involves personal risks of the kinds described, is inevitable.
>
> (Claxton 1991: 99)

Some types of explanations are more satisfying to our minds than others. Simpler ones, as a rule, win out over the more complicated: we will take the more direct of two equally good explanations – and may even overturn a slightly better but more complex one for a slightly worse but more straightforward one. And the more coherent, the more story-like and narrative-driven, the better – especially if it also explains a number of factors at once.

How are explanations part of the construction of meaning?

'Miss?'

'Yes, Terry?'

'When I wipe the window, why does my skin squeak on the glass?'

In 1983, Johnson-Laird (1983: 17) said 'understanding consists of having a "working model" of the phenomena in your mind'. 'Yes!' says the learner once the explanation, the working model, takes hold. 'That explains it'. Explanations have varied explanatory power. In classical philosophy, the explanatory product was known as the 'explanandum', that unit of explanation the made sense of an issue. In this chapter on school science, let me call it the 'chunk' of knowledge, the 'quantum of explanation', the *explandon*, it sounds more fitting as a basic science entity like hadron or lepton. When a teacher is in position to offer an explanation, then he or she is faced with a range of choices:

1 The size of the explandon

It is clearly inappropriate to provide a long, involved explanation when what is required is something crisp and succinct. The overly-long version runs the risk of information overload, of pouring in too much detail, of being off-putting and boring. A thirty minute treatise on the effects of electric current may not be an answer to why the light bulb has blown. Nor can the *explandon* be too curt or short: it then ceases to be an explanation at all, is insufficient, does not 'close the gap of understanding'. As in the Goldilocks Principle, the *explandon* must not be too big or too small, but must be just the right size to satisfy.

2 The level of the explandon

The same applies here. The teacher must weigh the audience, make judgements of suitability, and pitch the explanation at just the right level. There is no point in an A-level answer to a Key Stage 3 question, nor vice-versa. This is the real test of the teacher's skill, answering questions to non-science colleagues in the staffroom is vastly different to responding to questions at GCSE.

3 The 'fit' of the explandon

In the course of offering the explanation, teachers can generally see what effect their words are having, what reception is taking place, how well the *explandon* fits with the student's thinking, seems plausible, meets their need. Needless to say, offering an explanation of something is seldom a 'one-shot' affair, and the conversation and discussion that follows commonly allows for several attempts or greater elaboration, greater fit, to take place through, for example, the introduction of models, similes and analogies: 'It is like ...' 'Imagine that it ...' As we talk we are constantly monitoring our audience for their reception of the ideas.

4 The tone and texture of the explandon

It might be offered as a provocative explanation, slightly fun, clinically factual, deeply serious.

Explanations highlight incompleteness. An explainee might encounter a gap in understanding that may remain largely invisible until something happens to reveal the gap. A reasonable working analogy is that of a jig-saw puzzle. Explainers create for the explainee puzzle-solver a piece of information. They do so with the intention that it has the correct 'locking points' that allow it to fit into the explainee's 'gap of knowledge'. It must be the right size, no more or no less, the right colour and texture, and the 'lugs' on the puzzle-piece must fit exactly into the receiving whole, into the coherent scheme (as shown on the cover of the puzzle's box lid). We have all had the experience of forcing the wrong piece into the wrong part of the puzzle, having to work it free and then start over again with another *explandon*.

So things commonly go wrong, of course. For example, an explanation of how something works will fail if it provides too much detail, if it presupposes too much knowledge or skips over essential details. As explainers, we try to adjust our explanation to take into account the other person but, by using our own knowledge as the starting point, we may be guilty of distorting it in terms of our own needs. 'I don't understand why you don't understand!' 'Any idiot could see that's the case!' 'It's so obvious as to be untrue!' It is a process that influences explanations and makes us miscalculate the informational common ground between explainer and explainee (Clark 1996). Although, as Clark says, speakers attempt to negotiate a common ground the, explainer's egocentric bias intrudes.

Explanations can also be seen to be bad if they fail to cohere or 'hang together.' The different elements of an explanation must work in concert with each other to achieve an internally-consistent package. The doctrine of *coherentism* is seen as an important alternative to *foundationalist* views of scientific explanations that attempt to reduce all phenomena to the bedrock of physics (Amini 2003). Coherentism argues that a set of statements at a particular level of explanation, such as the psychological, can cohere as a tightly organised and interrelated unit that offers insights and explanations in its own right, without having to depend on lower levels of explanation (Amini 2003).

That said, one key drawback to the jigsaw analogy is that the picture on the puzzle box remains static – it is *the* picture, beginning and end. The explainee's picture, of course, changes, and that is one major points of an explanation: its value as it changes the receiver's picture. In that sense, an explanation is whatever answer that yields relevant information as a response to whatever kind of question. The explanatory force has little to do with truth but a lot to do with making sense to the questioner, and enabling him or her to understand better and 'change their view'. Gestalt psychologist Fritz Heider put it this way:

> If I find sand on my desk, I shall want to find out the underlying reason for this circumstance. I make this inquiry not because of idle curiosity, but because only if I refer this relatively insignificant offshoot event to an under-lying core event will I attain a stable environment and have the possibility of controlling it.
>
> (Heider 1958)

What is good about self-explaining?

'Sir?'

'Yes Jac?'

'Why do cuts heal?'

Let me move to another part of the process of explaining, called self-explaining. Self-explanation is 'the activity of explaining to oneself in an attempt to make sense of new information, either presented in a text or in some other medium' (Chi 2000: 164). That is, where a student attempts to explain something to herself – as opposed to having it explained to her, or even to her explaining it to another person (Roy and Chi 2005). It is defined as the sense-making process that an individual uses to reach a greater understanding of something. In school science, this 'something' could be what the teacher has said, material like a worksheet, textbooks, worked-out examples, diagrams or other multimedia materials. In my view, it is a generative process that provides a direct link between the activity of constructing meaning, and that of reorganising knowledge. It is a form of reflecting and elaborating on one's own thinking. A substantial number of studies have reported that students do develop a deeper understanding of material through this form of explanation as they generate explanations to themselves while learning (Aleven and Koedinger 2002). These studies show, for example, that the constructive activity of self-explanation is more beneficial for learning than, say, the activity of simply reading, repeating, or paraphrasing the material to be learned, which does not encourage generating inferences beyond the given information.

Noreen Webb's research has focused on the role of exchanging explanations – elaborating on an answer as opposed to merely sharing it. First, she argues that the act of explaining correlates with increased achievement (Webb 1982b, 1991). This is a positive act: the more we can get young people to engage in explanations to themselves and to others, then the more they can achieve. Second, receiving an explanation from someone else, on the other hand, has shown a weak and inconsistent correlation with achievement (Webb 1982a, 1982b, 1991). Better to generate and construct your own explanation than have someone give it to you. Receiving accurate, understandable explanations is, naturally, more helpful than errant or confusing ones (Webb 1991). Third, vague requests for help do not correlate with achievement, possibly because they tend to elicit low-level responses such as merely supplying the answer. More specific and persistent requests for help tend to elicit more substantial explanations which drive toward conceptual understanding (Webb and Mastergeorge 2003a, 2003b). So, the better the question, the more likely is a better explanation.

The mechanisms of self-explaining are only partially understood. Some students have a natural tendency to self-explain, while other students do little more than read out or repeat the content of the example or text. Most broadly, self-explaining promotes learning by requiring the integration of new information with that which is already established: it guides students to make more internal and external connections. In the jigsaw analogy, the person solving the puzzle is the one who creates and shapes the piece, the *explandon*, and then fits it into place. For example, in self-explaining the solution to a physics problem, it might become apparent that a solution is appropriate only because of the relationship between forces (which figure in the explanation) and not the more superficial characteristics of the problem (which do not). Explaining this to oneself seems to make it easier to generalise and to transfer problems by isolating relevant issues set out in the problem. That is, it is the *process* of self-explaining rather than the actual *explandon* itself that matters. It is the process of generating the content that is more important for learning than the content itself. Students learn more when they *generate* the explanation for themselves rather than merely read and *comprehend* someone else's explanations. The learning that results from self-explanation is largely due to the active generation of the missing information in the form of justifications for the application of each problem-solving step.

Another element is that, because self-explaining is often carried out in a continuous, ongoing and rather piecemeal fashion, it gives rise to multiple opportunities for the learner to build up and revise his or her internal representation that, ultimately, may lead to an integrated whole, a fuller picture. When completing a jigsaw puzzle, people commonly pick up and try out pieces, not quite at random, but according to approximate shape and colour. In this way, self-explaining forces learners to explore their understanding, and to generate inferences – which kinds of pieces fit and which don't – and this helps them to monitor and revise their knowledge (Renkl and Atkinson 2002). These active learning mechanisms may push learners to construct a more integrated representation (Chi *et al.* 1994). According to Renkl and Atkinson (2003), self-explanation activities are effective because self-explaining contributes directly to the construction of a learner's mental models of how things work: how, in biology, the heart works in pumping blood, how an electrical circuit works, what drives osmosis.

Conclusion

Learning with understanding in science involves the usefulness of possible explanatory ideas by using them to make predictions or to pose questions, collecting evidence to test the prediction or answer the questions and interpreting the result; in other words, using the science process skills.

(Harlen 1999: 130)

There is considerable diversity in kinds of explanations in terms of the causal patterns they invoke, the broad stances they employ, and the local domains in which they occur. That said, people throughout the world share the same drive for explanation, a similar assortment of explanatory styles and strategies for dealing with gaps, and similar developmental patterns. Our differences may lie in exactly which explanatory styles come to mind first in specific contexts, not in terms of fundamental explanatory abilities. The processes of constructing and understanding explanations are intrinsic to our cognitive and emotional lives from an early age, with some sense of explanatory insight present before children are even able to speak. Explanations are sought after and provided not only in interpersonal interactions but also within the mind of a single individual. Qualitatively different patterns of explanation seem to be used in talking about domains such as physical mechanics, biological function and social interactions.

The causal and relational complexity inherent in much of the world makes many explanations necessarily incomplete or flawed. In science teaching, we must therefore rely on coarser 'gists' that provide effective explanatory frameworks while nonetheless missing many details. We are adept at supplementing these gists often by 'outsourcing' knowledge to other experts (TV, textbooks, Internet, museums, field centres) and relying on the divisions of cognitive labour that occur in all cultures. Overall, we want our students to use a wide variety of ways to recognise poor, or plainly bad explanations, and to be able to choose among competing explanations that may sound reasonable. These skills, however, need to be taught.

Key questions

1   What can't be explained? It is certainly true that it is hard to explain something to someone who doesn't want that explanation (Darwinism to a creationist; a spherical world to a flat-Earthist), but are there things – in and out of science – that cannot be explained? What would be your list?

2   How can we improve explanations? I guess no one has a monopoly on wisdom, but there are certainly some people who have the 'gift', who can see through issues with sufficient insight to be able to illuminate it easily. But is that a gift, or can all teachers generate their own explanatory systems so that they develop their own insights?

Further reading

Shank. R. (2011) *Teaching minds.* New York: Teachers College Press. I really like all of Roger Shank's writing, and this is one of his recent books. Real food for thought.

Newton, D. P. (2000) *Teaching for understanding: What it is and how to do it.* London: Routledge. One of the best books on the subject. It can be dense in parts but well worth the effort.

Sheldrake, R. (1994) *Seven experiments that could change the world.* London: Fourth Estate. Okay, so have a go at explaining some of these issues to someone else – especially a sceptical scientist!

References

Aleven, V., and Koedinger, K. R. (2002) An effective metacognitive strategy: Learning by doing and explaining with a computer-based Cognitive Tutor. *Cognitive Science,* 26(2): 147–79.

Amini M. (2003) Has foundationalism failed? A critical review of *Coherence in Thought and Action* by Paul Thagard. *Human Nature Review,* 3: 119–23.

Cartney, P., and Rouse, A. (2006) The emotional impact of learning in small groups: Highlighting the impact on student progression and retention. *Teaching in Higher Education,* 11(1): 79–91.

Chi, M. T. H. (2000) Self-explaining expository texts: The dual processes of generating inferences and repairing mental models. In R. Glaser (ed.) *Advances in instructional psychology.* Mahwah, NJ: Erlbaum (pp. 161–237).

Chi, M. T. H., de Leeuw, N., Chiu, M. H., and LaVancher, C. (1994) Eliciting self-explanations improves understanding. *Cognitive Science,* 18: 439–77.

144   Mike Watts

Clark, H. H. (1996) *Using language*. London: Cambridge University Press.

Claxton, G. (1991) *Educating the inquiring mind*. London: Harvester Wheatsheaf.

Harlen, W. (1999) Purposes and procedures for assessing science process skills. *Assessment in Education*, 6(1): 129–44.

Heald, T. (ed.) (2006) *Conversations with Carl Sagan*. Jackson: University Press of Mississippi.

Heider, F. (1958) *The psychology of interpersonal relations*. New York: Wiley.

Johnson-Laird, P. N. (1983) *Mental models*. Cambridge, MA: Harvard University Press.

Lombrozo, T. (2010) The structure and function of explanations. *TRENDS in Cognitive Sciences*, 10(10): 464–70.

McDermott, L. C. and Shaffer, P. S. (1992) Research as a guide for curriculum development: An example from introductory electricity, Part I: Investigation of student understanding. *American Journal of Physics*, 60: 994.

Newton, D. P. (2000) *Teaching for understanding: What it is and how to do it*. London: Routledge.

Piaget, J. (1929) *The child's conception of the world*. London: Routledge and Kegan Paul.

Piaget, J. (1971) *Biology and knowledge*. Chicago: Chicago University Press.

Renkl, A., and Atkinson, R. K. (2002) Learning from examples: Fostering self-explanations in computer-based learning environments. *Interactive Learning Environments*, 10: 105–19.

Renkl, A., and Atkinson, R. K. (2003) Structuring the transition from example study to problem solving in cognitive skills acquisition: A cognitive load perspective. *Educational Psychologist*, 38: 15–22.

Roy, M., and Chi, M. T. H. (2005) The self-explanation principle in multimedia learning. In R. E. Mayer (ed.) *The Cambridge handbook of multimedia learning*. Cambridge: Cambridge University Press (pp. 271–86).

Strevens, M. (2009) *Depth: An account of scientific explanation*. Cambridge, MA: Harvard University Press.

Webb, N. M. (1982a) Group composition, group interaction, and achievement in cooperative small groups. *Journal of Educational Psychology*, 74(4): 475–84.

Webb, N. M. (1982b) Peer interaction and learning in cooperative small groups. *Journal of Educational Psychology*, 74(5): 642–55.

Webb, N. M. (1991) Task-related verbal interaction and mathematics learning in small groups. *Journal for Research in Mathematics Education*, 22(5): 366–89.

Webb, N. M., and Mastergeorge, A. (2003a) Promoting effective helping behavior in peer-directed groups. *International Journal of Educational Research*, 39(1): 73–97.

Webb, N. M., and Mastergeorge, A. M. (2003b) The development of students' helping behavior and learning in peer-directed small groups. *Cognition and Instruction*, 21(4): 361–428.

Webb, N. M., Troper, J. D., and Fall, R. (1995) Constructive activity and learning in collaborative small groups. *Journal of Educational Psychology*, 87(3): 406–23.

Chapter 11

Science education through contexts: is it worth the effort?

John K. Gilbert

Introduction

New aims, content and approaches to the teaching and learning of science spring up from time to time. In recent years, examples have included inquiry-based learning and computer-managed instruction. It seems that, as many such innovations evolve from bright ideas, through being included in national curricula, introduced into teacher professional development activities, and implemented in classrooms and laboratories, they progressively lose their distinctive natures (Van Den Akker 1998). The problem seems to be that too many innovations are adopted without their detailed implications for teaching, learning and assessment being thought through beforehand. In the long run, the consequence is that nothing much really changes as a result of many, if not most, innovations. As a consequence, the view of science education as the didactic exposition and regurgitation of 'facts', to a large extent, prevails.

One innovation that has come into prominence in recent years is that of 'the context-based teaching and learning of science'. School science departments that wish to be seen as 'up-to-date' may well be encouraged by their managers to adopt this approach for the development and implementation of schemes of work. This chapter identifies some of the key issues that have to be addressed in taking a decision on whether to adopt a context-based approach to learning in science and, if it is decided to so do, what is involved in carrying it through with some hope of success.

What is a 'context' in science education?

Let's get the most difficult bit over with first: the meaning of 'context'. The word 'context' refers to the totality of a setting in which a concept can be encountered. That concept may be of an inherently abstract idea (e.g. force), of an object (e.g. a car), of a process (e.g. fractional distillation), or of an event (e.g. a tsunami). A context provides a description of the circumstances in which a concept is used, such that the meaning intended can be fully understood. In terms of the learning of concepts, the function of a context is thus to facilitate the acquisition of a coherent meaning for a concept by placing it within a broader perspective. A given context can usually only be fully understood by the use of several, if not many, concepts. Every context of educational interest is a manifestation of the culture of a particular person, group, organisation, nation or historical period. For example, the context provided by the 30th Olympiad in London provided the setting in which the sum of the concepts underlying the events within the Women's Heptathlon Final made sense.

To be of value as a basis for systematic education, a context must meet four criteria (Duranti and Goodwin 1992). First, the setting chosen must be one in which students will/might/should take both an interest in and be likely to understand. Second, it must be capable of facilitating that high level of interaction between the teacher and the students needed to ensure that learning takes place. Third, it must enable a clear understanding of specialist language to be introduced, defined and used in the educational process; in short, it must deal concisely with the new concepts that are being introduced. Fourth, it must relate to, use, and build on the students' relevant prior knowledge, so that they acquire a steadily widening appreciation of the world-as-experienced. The justifications for seeking to meet these four criteria in science education, together with exemplars of the conditions in which they might be met, are given later in this chapter.

Why base science education on learning from contexts?

The provision of an effective science education is being encouraged by governments throughout the world for a variety of reasons, the supposed link to economic development and greater prosperity perhaps being the most heavily promoted. Despite this, the uptake of science courses when they are no longer obligatory – usually after the attainment of the 'school leaving age' – often falls short of what is thought economically desirable, except in periods of economic depression. This lack of enthusiasm on the part of students is put down to an inter-play of five reasons that, it is currently being suggested, could be tackled – at least in part – by the adoption of context-based science courses (Gilbert 2006).

The first reason lies in the fact that the science curriculum is undoubtedly overloaded with content (Rutherford and Ahlgren 1990). This means that individual concepts, even the most important ones, are tackled too superficially in classes for deep understandings to be acquired. Attempts to 'thin down' curriculum prescriptions usually end in failure, for every item in an existing syllabus will have its strong advocates. The specification of particular, carefully chosen contexts can automatically reduce the curriculum load by focusing attention on key concepts, and thus evading the historical detritus with which existing curricula are cluttered.

The second reason is one consequence of this overloading. It leads to information often being learned in isolation, with little likelihood of integrated mental schema being formed by students between the concepts involved. This fragmentation of knowledge does not encourage good student engagement in classes, and hence leads to the ready forgetting of what is taught. A given context invokes the use of a finite number of concepts, and the emphasis is on using them to forge an overall appreciation of that context. Thus the teaching of context-based courses should lead to the formation of coherent mental schema that link the new concepts both together and to existing concepts.

Another consequence of the overloading of syllabuses and the acquisition of fragmented knowledge is that the transfer of knowledge to other, very different contexts does often not take place. Concepts that are learned using idealised or exemplary examples are too often not transferable to other, particularly to everyday, examples (Osborne and Collins 2000). For example, the teaching of Newton's Laws of Motion without reference to the impact of inter-surface friction in everyday life produces knowledge that is only applicable in outer space. The third reason for the teaching of concepts within the framework of specific contexts, coupled to an appreciation of those contexts overall, is thus said to be that it enhances the ability of students to understand other, only somewhat similar, contexts. The Queensland Studies Authority view this as the most important reason why all science, and in particular chemistry, should be approached through the examination of contexts (Queensland Studies Authority 2004).

The use of idealised or exemplary examples as the basis for all teaching too often leads students to see little, if any, relevance of the concepts being taught to everyday life and to problems in which they are interested. As a fourth reason for their use, the careful selection of contexts, in which students have (or can be persuaded to have!) an interest, should lead to improved engagement in learning and subsequently to a more ready transfer of knowledge.

The fifth reason lies in the tensions between the demands of providing a 'science education for possible future scientists' and a 'science education for everyday life'. These tensions have become progressively greater as societies have placed greater emphasis on the latter whilst seeking to preserve the former (Bybee 1997). The careful selection of contexts that either simultaneously or separately address the aims of these different curricular emphases can accommodate these tensions. Moreover, context-based courses, for whatever purposes they are adopted, are claimed to lead to improved scientific reasoning skills, for example argumentation, when taught through project work (Krajcik *et al.* 2008).

These five issues may well map reasonably well on to the educational circumstances and aspirations of a particular school. If so, the next step must be ensure that the learning that results from context-based schemes of work is compatible with the high-stakes public examinations to which the school is committed.

How can context-based teaching be introduced?

Once it is clear that context-based work, if successful, might lead to improved attitudes and learning – together with better examination results – in a particular school, the major task is to decide how to introduce the approach. Many teachers will want to do this gradually. The most cautious way of making such a change is to insert detailed published context-based lesson plans into existing schemes of work as replacements for conventional approaches. The 'Science and Technology in Society' materials (ASE 1986) provide a wealth of diverse, well-trialled examples, each of which take only a few lessons to implement. A somewhat less cautious approach is to augment the conventional teaching of key concepts with the substantial use of the materials from a published course that place the main focus on the application of those concepts. A good example is the 'Science for Public Understanding' course (Hunt and Millar 2000). Once teachers (and the students!) are comfortable with the different approach to teaching and learning that context-based work entails, it may be timely to adopt an entire published examination-preparation course. In the United Kingdom, the suite of 'Salters' courses, e.g. 'Salters Nuffield Advanced Chemistry' (Burton *et al.* 1994), and in the Netherlands, the 'Dutch Physics Curriculum Development Project' (PLON 1988), are well-tried examples. Once teachers have extensive experience in using pre-published context-based courses, they may wish to develop materials that relate to local contexts of particular interest to their students. This, of course, requires extensive commitment of time, resources and expertise, preferably in a framework built around a group of like-minded colleagues. However, this investment can be very worthwhile, for experience in the Netherlands shows that it can be both professionally rewarding in itself and lead to a sense of ownership of the product (Wieringa *et al.* 2011).

There are several types of context that may prove capable of sustaining the ambitions of the genre. If a context can be seen by students as providing insights into themes or problems of personal relevance, then it may be a strong candidate as the basis for teaching (Campbell *et al.* 2000). In particular, a context that focuses on the immediate circumstances of students, whether personal (Taasoobshirazi and Carr 2008) or social (Sadler 2009), may provide suitable educational opportunities. Of these immediate circumstances, examples that relate to students' possible or actual career aspirations may be fertile sources of ideas for teaching materials (Aikenhead 2007). However, while it is important to select contexts that seem likely to engage the sustained interest and engagement of students, it is equally important that the teaching be both appropriately organised and then take place in a suitable manner.

How might a curriculum based on contexts be organised?

Looking at the tactics for introducing context-based teaching outlined above suggests that four models of strategy can underpin attempts to do so. These draw on particular traditions in science education (Gilbert *et al.* 2011). Each of these models can be evaluated against the four criteria for context-based teaching set out earlier. These criteria are whether a context is being used to fully or partially provide all or some of the following:

1   A systematic introduction to the social, temporal, or spatial framework of a community of practice of some kind that defines it.
2   Opportunities for systematic and sustained interaction between the teacher and the students.
3   Opportunities to become very familiar with the meanings of the specialist concepts that are introduced.
4   Careful and detailed linkage to students' relevant prior knowledge.

We consider each model in turn.

Model 1: Contexts provide the application of concepts after formal teaching

This is the most conservative approach. In it, key concepts are taught in a traditional way, being introduced separately by use of abstract and/or idealised situations. An example is the teaching of the behaviour of lenses by reference to contrived 'experiments' in the use of the optical bench. Because such a treatment will only produce the active engagement of those students who are the most committed to the study of science, more convincing applications are introduced after the formal teaching is finished, for example, in order to explain the rationale behind the operation of telescopes. Such applications will not be the basis for the formal assessment of learning. As a consequence they will be seen by many students as mere decoration in a curriculum of abstract concepts. Model 1 does not meet any of the criteria for context-based teaching to any acceptable extent. Thus: concepts are taught separately; the teaching is likely to be didactic in style, so that little two-way teacher/student interaction takes place; familiarity of students with the concepts is constrained by the contrived situations in which they are taught; there is little especial need for the teacher to find out how the new concepts link to what students already know.

This model is likely to be adopted by teachers who are either unaware of the rationale for context-based teaching or who do not believe such an approach to be valuable.

Model 2: Contexts provide applications during formal teaching

In this model, the teaching of concepts remains both direct and related to idealised exemplars. However, the applications of those concepts in realistic contexts are introduced both as a framework for the teaching and are referred to continually throughout. In terms of the four criteria, the status of contexts is apparently heightened, but that of the other three criteria remain as for

Model 1. Model 2 has two additional weaknesses. First, it is very likely that much greater emphasis will be placed on the concepts themselves in the reality of the classroom, thus implicitly lowering the significance of the contexts as such. Second, as (Layton 1993) has pointed out, the meaning of concepts used in science and in its applications are often different. For example, in science the concept of energy is explained in terms of the 'kinetic properties of atoms, ions and molecules', in technological applications of the concept, the meaning adopted is that of 'fluid flow'. Thus shifting from a scientific to a technological frame of reference may merely confuse students over the meaning of a concept.

This model is likely to be adopted by teachers who still place great reliance on traditional teaching methods, but who also believe that the application of the concepts taught are important, if only for motivation purposes.

Model 3: Contexts provide a framework for personal mental activity

In this Model, criterion 1 for context-based learning is fully met, but the task for teaching is seen to be that of stimulating mental activity by each student as an individual. This approach is based on 'personal construct psychology' (Pope and Keen 1981), which assumes that a person's learning is entirely due to the individual, internal, mental, activity of that person. This means that the major value of a teacher for many individual students will be on those (inevitably rare) occasions when one-to-one interactions take place. It will not be entirely clear to either the teacher or the individual student both whether specialist language (the 'carrier wave' for new concepts) has been fully understood and whether linkages have been made between existing and new knowledge.

This model will be adopted by teachers who believe that learning is entirely a personal matter, this view being perhaps common because of a continued emphasis on Piaget's ideas in teacher training.

Model 4: Contexts as providing a range of social circumstances

In this model, there are two meanings for 'social circumstances'. The first of these refers to the contexts that are fully or partially addressed as particular concepts are used to provide detailed explanations of them. The second refers to the way that teaching and learning are conducted. In this model, the teacher and a class of students work collaboratively in what has been termed a 'community of practice' to explore contexts, leading to 'situated learning' (Greeno 1998). The primary focus on the contexts and the full implications of teacher-student and student-student interactivity suggest that all the four criteria for context-based learning can be fully met.

In its purest form, Model 4 assumes that a teacher has an excellent understanding of the concepts to be addressed and of the contexts in which this is to take place. Equally importantly, a teacher must both believe in the efficacy of 'situated learning' and have a relationship with students such that a collaborative approach to work is realistic.

The provision of any approach to teaching will only lead to effective learning if the teacher actively believes in the principles of teaching and learning that underlie it (Van Driel *et al.* 2005). Model 4 is the one that most fully realises the potential of context-based learning. However, it is important that an individual teacher adopts a model that is as closely congruent with their current beliefs as possible. In doing so, they will acquire experience such that they feel able to enter into suitable professional development activities (Gilbert 2009) that lead them towards the adoption of Model 4.

What does context-based classroom teaching involve?

With some experience in the classroom, every teacher develops a complex and very personal style for their work. It is only possible here to point to some of the major elements in such a style: those that are particularly relevant to context-based teaching. These are the particular strategies and tactics adopted for: the use of language in the classroom; ensuring that the teaching of individual concepts is effective; and facilitating a high level of student engagement though 'inquiry-based' work. We again consider each of these elements in turn:

Language in the classroom

The styles of language use in the classroom have been described by Mortimer and Scott using four terms: authoritarian, dialogic, non-interactive, interactive (Mortimer and Scott 2003: 35). In authoritarian language use, the teacher acts as the sole possessor of the knowledge that is to be imparted to the students, the latter taking a docile, recipient roles. Dialogical language use, on the other hand, is more subtle: while (the students must hope that!) the teacher knows more about the subject being taught, the students are able to put their ideas forward, acceptable meanings emerging from the evaluation of all the ideas that have been put forward. In non-interactive language use, the teacher speaks for most if not almost all of the class time. However, in interactive language use, both the teacher and the students ask questions, make statements and argue, these then being discussed by all.

In a Model 4 approach to context-based learning, the dialogic and interactive patterns of language use predominate. This high level of teacher-student and student-student language use ensures (as far as is ever possible!) that all are actively engaged and are therefore learning. Specialist language is used and revisited frequently, thus leading to a higher likelihood of a deep understanding being acquired, while new knowledge is mentally linked to existing knowledge. This approach is effective when the teacher knows the subject thoroughly, when the students accept this, and when classroom discipline is good with a high level of on-task activity.

The effective teaching of concepts

The effective teaching and learning of individual concepts in science education has been thoroughly research, such that clear advice on procedures is readily available (e.g. Scott *et al.* 2007). However, a Mode 4 approach to context-based teaching presents some additional requirements. First, there is a need for the teacher to identify the concepts that are to be invoked in the exploration of the contexts selected for study. This will be different from the traditional approach, where concepts have a prescribed place in the curriculum of subjects e.g. in physics. Second, as it is the exploration of the context that is the organising principle of the teaching, the teaching of those concepts has to take place on a 'just in time' basis (Bennett 2003: 110). This means that only those aspects or manifestations of a concept will be acquired that are needed for the explanation of some aspect of the context. To ensure that a comprehensive understanding of a concept is acquired, the teacher will adopt a 'spiral curriculum' approach, in which it is revisited within several contexts.

Inquiry-based learning

A central tenet of Model 4 context-based teaching is that the students should be actively engaged in acquiring knowledge. This means that they search for and evaluate documents, pose hypotheses, make predictions and test them out, both in the laboratory and with the use of data banks. This activity-centred approach to science education is known as 'inquiry-based learning' and has been the subject of extensive research and development work (e.g. Lederman 2007). In order to provide effective context-based teaching, a teacher will need to be 'comfortable' with the design and management of inquiry-based teaching.

What major problem might students have in learning from contexts?

Students have a range of problems in learning science, of which those that stem from their existing 'alternative conceptions' of the ideas being discussed have been found to be important (Gilbert and Watts 1983). Avoiding and remediating these is possible (Scott *et al.* 2007).

One of the major attributes claimed for context-based learning is that is promotes the ability of students to use concepts learned in one context in a different context, i.e. to engage in 'transfer of learning'. The knowledge and skills involved are not emphasised in conventional (i.e. authoritarian and non-interactive) teaching, which can explain a general inability to do so by students (De Corte 2003). Unless this problem is directly addressed, a major attribute of context-based teaching will not be realised.

'Transfer' can be divided, if only for the purposes of study, into three types (Gentner 1983):

1 *Near transfer* – In near transfer, the concepts involved in learning about a new context are very similar to those already experienced. For example, the concepts involved in the study of the contexts of how ice cream, yoghurt and cheese are made are sufficiently close that the study of one of them could readily lead to an understanding of any of the others.

2 *Further transfer* – In further transfer, the concepts involved in the study of the contexts of, say, how various antibiotics are made are somewhat different, although the processes undertaken differ only in detail. For example, the production of penicillin to that of aureomycin is sufficiently similar for the latter to be understood after the study of the former.

3 *Far transfer* – In far transfer, the concepts used to understand one context must be considerably modified if understanding of an apparently very different context is to be achieved. For example, the study of food technology can lead on to the study of biochemistry, although the concepts involved would have to be considerably modified to do so (Gilbert *et al.* 2011: 830–1).

These three types of transfer are linked in a continuum by the use of analogy (Hesse 1966), the mechanism of which Gentner has examined (Gentner 1983).

The implications of this analysis are that, if context-based teaching is to lead to students being able transfer concepts between contexts, of whatever 'distance apart', then they must have both a theoretical understanding and extensive practical experience of the operation of analogy. This seems to be rarely taught in school science, even though ways in which it can be done have been long established (Glynn 1991).

How can a teacher be sure that the study of contexts is worthwhile?

Any teacher who has invested a great deal of professional time in adopting/adapting/implementing context-based teaching will want reassurance that that the effort has been worthwhile. Particularly when Model 4 has been implemented, the ebb and flow of ideas and opinions in a classroom will enable a constant monitoring of engagement, knowledge and motivation to take place. The outcome of public examinations will provide an important measure of the impact on knowledge acquisition and use, but the information obtained will only be valid if the assessment techniques used are commensurate with the aims of the teaching approach adopted.

The outcomes of research studies (which look at a wider range of outcomes than do public examinations) ought to give teachers some measure of what they might be able to achieve. Here are some key findings, albeit from a slender body of studies, on core questions:

What impact do such courses have on students' engagement in science lessons?

The impact has been found to be positive (Ramsden 1997), particularly where the contexts were seen as directly relevant to their lives (Campbell *et al.* 2000).

What impact do such courses have on students' intention to continue with the study of science?

Little significance for the direction of future study has been found (Ramsden 1997), although it is possible that the students surveyed had too brief an exposure to such courses for any impact to have been felt. It must, of course, be recognised that a wide variety of factors influence students' future choice of course, for example general economic circumstances, family traditions, peer opinion and schools' capabilities.

What impact do such courses have on students' attainment in examinations?

The results here are ambiguous. There is evidence both that achievement can be improved relative to conventional teaching (Gutwill-Wise 2001) and, on the other hand, that it can have no significant effect (Taasoobshirazi and Carr 2008). Given that the quality of the teaching provided in such courses may vary considerably, perhaps dependent on which Model and particular contexts are adopted, this ambiguity is to be expected.

The overall conclusion to be drawn is perhaps that context-based work does improve the 'atmosphere' of science classrooms, while – entirely reasonably – consistent and sustained effort is needed to achieve the potential of the approach.

What factors influence a decision about context-based teaching?

There are a number of pragmatic factors concerning the current teaching and learning of science in a particular school that may lead to a decision to introduce context-based learning. These factors are the extent to which:

- the present teaching schemes are too overloaded to allow effective teaching and quality learning to take place;
- students lack engagement with their science classes, acquire fragmented knowledge, and cannot transfer what they learn to contexts other than those initially experienced;
- too few students opt for science courses when they are able to exercise free choice.

There may be broader issues that underlie these pragmatic factors. For example:

- There are real problems in making curricular provision simultaneously of both 'science for all' and 'science for potential future scientists'.
- Whether there is a public examination available for which context-based teaching is an appropriate preparation.
- Whether established approaches to the teaching of science in the school make use of classroom language and inquiry-based learning that are commensurate with any model of context-based teaching.

Conclusion

It does seem that context-based science teaching can lead to an improvement in the quality and consequences of the learning achieved. However, if it is decided to adopt this approach, then it is highly desirable that this introduction is made gradually, giving both the teachers and the students time to adapt to the changes involved. If this is done, there is every chance that a real and valued innovation will be both made and sustained.

This chapter has outlined the main components involved in, and challenges associated with, such a gradualist approach. As the innovation is progressively introduced, the nature of contexts that are of interest to, and which can be readily understood by, students will emerge. The students will gradually develop greater skills of interactivity, learn how to clarify their ideas, and learn how to relate new information to their existing knowledge. In addition to supporting the students in their evolving skills, the teachers involved will gradually learn more how about to accommodate differing curricular aims, how to differentiate between the major and minor concepts of science, and how to facilitate the transfer of knowledge.

Adopting science education through contexts is worth the effort, for both teachers and students, but only if approached carefully and systematically.

Key questions

Having read this chapter, you might think of answers to the following questions:

1   Can you think of a particular context that would enable you, by implementing a Model 4 context-based course, to teach a group of concepts to a particular group of students?
2   What actions would you need to take in order to feel able to adopt that context-based approach?
3   Having taught your context-based course, how would you ensure that students transferred their learning to other contexts?

This chapter takes a broad perspective on the ways in which the ideas of science may be communicated and of the actors and approaches involved in that process. It offers different ways of viewing any context within which you might structure your teaching. You might then approach any given context by asking: What might the purposes of the teaching be? To whom might the learning be of use and for what purposes? From whose perspective might the teaching be constructed?

Further reading

Gilbert, J. K., Bulte, A. M. W., and Pilot, A. (2011) Concept development and transfer in context-based education. *International Journal of Science Education*, 33(6): 817–37. This article explores the ideas contained in this chapter in greater depth and could form the background to any research assignment that you undertake.

Stocklmayer, S. (2013) Engagement with science: Models of science communication. In J. K. Gilbert and S. Stocklmayer (eds) *Communication and engagement with science and technology*. New York: Routledge (pp. 19–38).

References

Aikenhead, G. (2007) Humanistic perspectives in the science curriculum. In S. L. Abell (ed.) *Handbook of research on science education*. Mahwah, NJ: Lawrence Erlbaum (pp. 181–911).

ASE (1986) *Science and technology in society*. Hatfield, UK: Association for Science Education.

Bennett, J. (2003) *Teaching and learning science*. London, UK: Continuum.

Burton, W., Holman, J., Pilling, G., and Waddington, D. (1994) *Salters advanced chemistry*. Oxford: Heinemann.

Bybee, R. (1997) Towards an understanding of scientific literacy. In W. C. Graber (ed.) *Scientific literacy*. Kiel, Germany: IPN (pp. 37–68).

Campbell, B., Lubben, F., and Dlamini, Z. (2000) Learning science through contexts: Helping pupils make sense of everyday situations. *International Journal of Science Education*, 22: 239–52.

De Corte, E. (2003) Transfer as the productive use of acquired knowledge, skills, and motivations. *Current Directions in Psychological Science*, 12: 142–6.

Duranti, A., and Goodwin, C. (1992) *Rethinking context: Language as an interactive phenomenon*. Cambridge: Cambridge University Press.

Gentner, D. (1983) Structure mapping: A theoretical framework for analogy. *Cognitive Science*, 7(2): 155–70.

Gilbert, J. K. (2006) On the nature of 'context' in chemical education. *International Journal of Science Education*, 28(9): 957–76.

Gilbert, J. K. (2009) Supporting the development of effective science teachers. In J. Osborne (ed.) *Good practice in science teaching*. Maidenhead, UK: Open University Press (pp. 274–300).

Gilbert, J. K., and Watts, D. M. (1983) Conceptions, misconceptions, and alternative conceptions: Changing perspectives in science education. *Studies in Science Education*, 10: 61–98.

Gilbert, J. K., Bulte, A. M. W., and Pilot, A. (2011) Concept development and transfer in context-based science education. *International Journal of Science Education*, 33(6): 817–37.

Glynn, S. (1991) Explaining science concepts: A Teaching-with-Analogies model. In *The psychology of learning science*. Hillsdale, NY: Erlbaum (pp. 219–40).

Greeno, J. (1998) The situativity of knowing, learning and research. *American Psychologist*, 53(1): 5–26.

Gutwill-Wise, J. (2001) The impact of active and context-based learning in introductory chemistry courses: An early evaluation of the modular approach. *Journal of Chemical Education*, 77(5): 684–90.

Hesse, M. (1966) *Models and analogies in science*. Notre Dame University, IN: Notre Dame University Press.

Hunt, A., and Millar, R. (2000) *AS Science for public understanding*. Oxford: Heinemann Educational.

Krajcik, J., McNeill, K., and Reiser, B. (2008) Learning-goals-driven model: Curriculum meterials that align with national standards and incorporate project-based pedagogy. *Science Education*, 92(1): 1–32.

Layton, D. (1993) *Technology's challenge to science education*. Milton Keynes: Open University Press.

Lederman, N. (2007) Nature of science: Past, present, and future. In S. L. Abell (ed.) *Handbook of research on science education*. Mahwah, NJ: Lawrence Erlbaum (pp. 831–80).

Mortimer, E., and Scott, P. (2003) *Meaning making in secondary science classrooms*. Maidenhead, UK: Open University Press.

Osborne, J., and Collins, S. (2000) *Pupils' and parents' views of the school science curriculum*. London: King's College London.

PLON (1988) *Projekt Leerpakketontwikkeling Natuurkunde (PLON)*. Utrecht, The Netherlands: University of Utrecht.

Pope, M., and Keen, T. (1981) *Personal construct psychology in education*. London: Academic Press.

Queensland Studies Authority (2004) *Chemistry: Extended trial pilot syllabus*. Brisbane: Queensland Studies Authority.

Ramsden, J. (1997) How does a context-based approach influence understanding of key chemical ideas at 16+? *International Journal of Science Education*, 19(6): 697–710.

Rutherford, F., and Ahlgren, A. (1990) *Science for all Americans*. New York: Oxford University Press.

Sadler, T. (2009) Situated learning in science education: Socio-scientific issues as contexts for practice. *Studies in Science Education*, 45: 1–42.

Scott, P., Asoko, H., and Leach, J. (2007) Student conceptions and conceptual learning in science. In S. L. Abell (ed.) *Handbook of research on science education*. Mahwah, NJ: Lawrence Erlbaum (pp. 31–56).

Taasoobshirazi, G., and Carr, M. (2008) A review and critique of context-based physics instruction and assessment. *Educational Research Review*, 3(2): 155–67.

Van Den Akker, J. (1998) The science curriculum: Between ideals and outcomes. In B. J. Fraser (ed.) *International handbook of science education (Volume 1)*. Dordrecht, The Netherlands: Kluwer (pp. 421–48).

Van Driel, J., Bulte, A. M. W., and Verloop, N. (2005) The conceptions of chemistry teachers about teaching and learning in the context of a curriculum innovatiom. *International Journal of Science Education*, 27(3): 302–22.

Wieringa, N., Janssen, F., and Van Driel, J. (2011) Biology teachers designing context-based lessons for their classroom practice: The importance of rules-of-thumb. *International Journal of Science Education*, 33(17): 2437–62.

Part IV

Subject debates

Chapter 12

Are science teachers immune to reflective practice?

Richard Malthouse and Jodi Roffey-Barentsen

Introduction

This chapter discusses the apparent tension between science and reflective practice. The tension occurs when a scientist embarks on a training programme, or engages in professional development, in which reflective practice is an inherent part. We offer a new model of reflective practice referred to as Situated Reflective Practice (SRP) along with a range of strategies that assist reflective practice in groups. This moves the perspective of traditional reflective practice from a psychological arena into the sociological.

A scientific perspective

Emma is a chemistry teacher in the final stages of her Initial Teacher Training programme. Although class teaching can be challenging at times, she loves the interaction with pupils, especially her role in inspiring pupils to embrace the sciences – and chemistry in particular. She regards teaching as an opportunity to cascade her enthusiasm for her subject that is, on occasions, seen as 'difficult' or 'pointless' by her pupils. Emma's challenge is to turn these pupils' perceptions around, demonstrating to them the relevance of chemistry and making it achievable.

Emma's university lecturers face a similar challenge. Reflective practice has been an inherent part of education since the 1980s, regarded as fundamental to the professional development of teachers, based on the work of theorists who include Schön (1983), Kolb (1984), Boud *et al.* (1985), Kemmis (1985), Brookfield (1987) and Gibbs (1988). Our chemistry teacher Emma, however, is struggling with the concept of reflective practice.

Emma: I am a scientist. I don't do waffle and frilly stuff – I can't see the point. I think that, for scientists, everything is black and white, correct or wrong. Therefore, like in maths, there is no ambiguity in the answers, so there is little to reflect on what might have been or should have been.

She appears to be unable to relate reflective practice to herself and her own practice, and justifies this by stating that reflective practice is probably more something for those who make important decisions.

Emma: The level of science I work at isn't highbrow enough to be presenting new thoughts or theories. However, if I was a research scientist, then I could see the need for reflection/open discussion about a proposed theory. But, at the level I work at, there are no challenges to the theories, as they have been long established. Except for the possible discovery of the Higgs Boson and how that might impact on our current thinking.

Reflective writing tasks, in the form of critical incidents, self-evaluations and reflections on modules, form part of the assessment pattern of the programme. She was asked to rate her own reflective writing on a scale from one to four, based on a model adapted by Roffey-Barentsen and Malthouse (2009) from

Table 12.1 Levels of reflective writing

Level	Name	Characteristics
One	Descriptive writing	• Some references to emotional reactions but they are not explored and not related to behaviour. • Account relates to ideas or external information. • Little attempt to focus on particular issues. • Hardly reflective at all.
Two	Descriptive account with some reflection	• Descriptive with little addition of external ideas, some reference to alternative viewpoints. • Some notion of asking questions but no response, i.e. no analysis. • Sense of recognition that learning can be gained from the event.
Three	Reflective writing (1)	• There is description but it's focused with particular aspects accentuated. • Evidence of external ideas. • Some analysis. • Willingness to be critical of the action, self or others. • Some 'standing back' from the event. • Different perspectives are considered.
Four	Reflective writing (2)	• Description now only serves to set the context. • Clear 'standing back'. • Self-questioning evident, critical self-awareness. • Views and motives of others are taken into account, multiple perspectives. • Recognition that events exist in a historical or social context. • Observation that there is learning to be gained. • Recognition that personal frame of reference can change according to emotional state.

Source: Roffey-Barentsen and Malthouse (2009: 86).

Moon (2004). Level One entails being mainly descriptive and Level Four being critical and analytical, taking other perspectives in to consideration. Emma rates herself as Level One.

Emma: Of course scientists evaluate their work, but this is impersonal; teachers reflect, which *is* personal. I don't mind my work being criticised but nothing personal. I can talk it but I can't write it, I am a Level One.

Emma was not the only student who rated her writing so low; out of the 18 students in the class, another two also give themselves a 'Level One': a mathematician and a psychologist. All other students, who were not from a science-based subject specialism, felt confident their writing was at least at Level Two or Three. So, what happens to scientists? Why do they find reflective practice so hard to apply? Emma is fully aware of her barriers and would like to overcome them.

Emma: I think this is something I will need to reframe in my mind, as I have a 'block' against doing it – using the excuse as a scientist to not reflect. As a teacher I will have to, so I can improve my teaching standard.

Emma is not untypical, some science teachers experience difficulty embracing reflective practice; others are simply unwilling to engage with the concept altogether. In this chapter we argue that science teachers occupy two concurrent and contradictory paradigms. On the one hand their grounding in the sciences has embedded them firmly within a positivist paradigm. However, being a teacher introduces them to the need to embrace what some might view as a conflicting interpretative paradigm. Education, being a social science, lends itself towards methodologies that the scientist may see to lack rigour or even to be subjective in comparison to the natural sciences (Kind and Taber 2005). The science teacher is trapped between the two paradigms and struggles to come to terms with the need to embrace both. Whereas the positivist paradigm values concepts of objectivity, measurability, controllability, predictability and the constructs of laws, interpretative approaches emphasise the interpretation of phenomenon and the construction of meaning about particular observations. The subsequent conflict, the cognitive dissonance they experience, has been observed in various studies of pre-service science teachers (Loughran 2007; Paul 2010).

Teaching practice

It is not until a science teacher embarks upon the process of developing his or her teaching identity (Paul 2010; Gault 2011) that the tensions between the two paradigms emerge. According to Loughran (2007: 1046), 'What and how teachers learn in their teacher preparation programme is strongly influenced by their existing knowledge and beliefs; therefore, challenging these through creating dissonance is one way of generating opportunities for new learning'.

A study of pre-service teachers in Australia (Paul 2010: 4) observed a 'gradual development in professional reflective practice during their under-graduate training'. Notably, this study commented that this reflective practice 'largely reflected at the technical and factual levels' (ibid.). Olson and Finson (2009) support this comment while observing the reflective practices of ele-mentary science education students. Using Perry's 1970 four-stage model of intellectual development (Table 12.2), they scrutinised their end-of-semester reflective essays.

Of the 38 students participating in the study, 28 demonstrated dualistic thinking within their reflective practice journals (Olson and Finson 2009). The characteristics of the dualist approach are very much in line with that of the positivist paradigm. Positivism is based on Compte's assertion that '... all genuine knowledge is based on sense experience and can only be advanced by means of observation and experiment' (Cohen *et al.* 2000: 8). Considering a dualist/positivist perspective in respect of teaching practice is the belief that there is one right way to teach and everything else must therefore be wrong. Further, there exists an expectation that the science teacher will not only iden-tify and explain the '*right way* to teach, the *best* practice, and should be able to answer all of the questions' (ibid.: 48). Failure on the part of teachers who do not do offer the one correct answer are perceived by the dualist as 'fraud-ulent or not knowledgeable'.

This dualistic expectation that there ought to be one right way, one answer, one truth is wholly in keeping with the positivist paradigm, where the observer,

Table 12.2 Perry's 1970 model of intellectual development

Stage	Description
Dualism	Division of meaning into two realms – good versus bad, right versus wrong. All that is not success is failure. Right answers are to be memorized by hard work. Knowledge is quantitative. Agency is experienced as external, residing in authority, test scores, the right job.
Multiplicity	Diversity of opinion and values is recognized as legitimate in areas where right answers are not yet known. Opinions remain atomistic without pattern or system. No judgments are made among them, so 'everyone has a right to his own opinion; none can be called wrong'.
Relativism	Diversity of opinion, values and judgment derived from coherent sources, evidence, logic, systems and patterns allowing for analysis and comparison. Some opinions may be found worthless, while there will remain matters about which reasonable people will reasonably disagree. Knowledge is qualitative, dependent on context.
Commitment	An affirmation, choice or decision (career, values, politics, personal relationships) made in the awareness of relativism (distinct from commitments never questioned). Agency is experienced as within the individual with a fully internalized and coherent value structure.

Source: Olson and Finson (2009).

detached from the phenomenon, observes and reports dispassionately. The problem that occurs with this philosophical viewpoint is that, although dualism/ positivism may be appropriate for the study of natural science, it is far less useful within the arena of social science. Here there are often no right and wrong answers, just possibilities. Interpretivism in the various forms of phenomenology, ethnomethodology or symbolic interactionalism does not by its nature offer a 'yes' or 'no' response to a research question. But that is not the purpose; its purpose is to offer insight into a process rather than to identify an end result. Critics of the interpretivist paradigm maintain that it lacks verification and generalisation; they argue that subjective verbal reports and accounts can be incomplete or misleading (Cohen *et al.* 2000). Whatever the arguments for or against either of the paradigms, those who adopt the dualist/positivist stance may experience difficulty when dealing with social science issues. For example, authorities who present a more complex view of reality, rather than simply right or wrong, are often rejected by dualists as being inadequate or untrustworthy (Olson and Finson 2009). Rapaport notes further that:

> dualists panic when confronted with multiple solutions, compare and contrast tasks, and reflection tasks. Since one way is the right way, they feel they are wasting time when presenting other ways that must be wrong. They struggle to know which of the multiple ways presented is the right way.
> (quoted in Olson and Finson 2009: 48)

This has a direct bearing upon the dualist's ability to engage in activities associated with the interpretivist paradigm, and one obvious activity is that of reflective practice. Put bluntly, 'when dualists are asked to reflect, they cannot do it' (ibid.: 47).

McIntyre (1993) suggests that it is more appropriate to promote reflective practice with experienced teachers rather than novice teachers. The argument put forward is that, whereas the novice teacher gives a great deal of consideration to their teaching practice, the experienced teacher benefits from being intuitive, benefits from a larger repertoire of experience. It is further suggested that, whereas novice teachers are willing to incorporate ideas easily, experienced teachers find change more problematic. The inference here is that the experienced teachers will be more able to reflect and their professional practice will also benefit. This view is not universally shared, though, with many being in favour of encouraging reflective practice among novice teachers (Hyatt and Beigy 1999; Parsons and Stephenson 2005).

Perceptions

But what about the perceptions of the science student teachers? A study of science student teachers in Turkey (Efe 2009) considered students' attitudes in relation to reflective practice. The survey concluded (ibid.: 82) that:

- Experience and knowledge of teaching affect the way student teachers see the teaching profession.
- Science student teachers who are more involved in teaching practice and had more knowledge of teaching indicated a more positive view of reflective practice.
- Reflection requires practice; the more student teachers are encouraged to reflect on the ethical, social and political issues of educating pupils, the more their reflective skills will be enhanced.

It appears, therefore, that science teachers:

- are dualists;
- don't see the point of reflective practice;
- don't engage with reflective practice other than an account or facts;
- engage in reflective practice only once they begin their teaching practice.

The mantra of reflection and reflective practice (Ecclestone 1996) has been taken up by different kinds of profession for the education and development of their members (Zukas *et al.* 2010). It may even be seen as the 'sine qua non of professional development' (Edwards and Thomas 2010: 403). It is, however, not without its criticism. The notion that there can be no professional development without being a reflective practitioner can be questioned. Furthermore, it may be seen as a 'time-consuming extra', for which most professionals are too busy. Finlay (2008) identifies some areas of concern with regard to reflective practice, including ethical, professional and pedagogical concerns. First, there are ethical concerns. As part of their coursework, trainee teachers are asked to write reflective essays, to keep a journal and generally engage in reflective practice. They do not have any choice in this, but have to conform to what is demanded of them by those in authority (Quinn 2000). They are asked to share their feelings, values and beliefs, otherwise, it is argued, they cannot develop their professionalism. This raises the issue of confidentiality. Should students be asked to share these reflections with their tutors or others within the department? Furthermore, it is generally assumed that reflection is good and beneficial, however, being a reflective practitioner may not always be a positive experience. It could be that, especially for those who are new to the concept (such as new or trainee teachers), the focus is on the negative, on things which have not gone so well within their teaching. The constant notion of not being good enough could result in a depressed state of mind, with consequences for the teaching profession which might lose conscientious teachers as a result. The second, the professional concern, focuses on what may happen when reflective practice is conducted uncritically. Finlay (2008) warns that on such occasions existing prejudices can be reinforced and potential bad practice within the school or department continued. It is confirmed that 'unless teachers develop the practice of critical reflection, they are strapped in unexamined judgments, interpretations, assumptions and expectations' (Larrivee 2000: 293). This notion may be further complicated, as teachers bring with them their tacit or implicit knowledge. These terms have been described by Argyris and Schön (1992: 10) as 'we know more than we can tell and more than our behaviour consistently shows'. They give the example of tacit knowledge where we recognise a face from thousands without being able to say how we do so, or, alternatively, do something without being able to fully explain why. Teachers may not even be aware of their tacit knowledge of the theories-in-use (Kinsella 2007) which shape their practice; they do not recognise that there are choices they can make, making changes to the established routines.

Furthermore, the emphasis on individual teachers being responsible for improving practice may contribute to relieving an organisation from doing so (Quinn 2000). The pedagogical concerns raised by Finlay (2008) include issues such as the readiness of students to reflect. She argues that those new to reflective practice are inclined to use mechanical, prescribed models, following these blindly and unquestioningly. They can only reflect on a limited number of experiences and they do not yet possess the depth of knowledge that is required to make a significant difference. Therefore, the reflections can be superficial and possibly ineffective. In turn, this may confirm to those sceptical about reflective practice that they really are wasting their time. A main limitation of reflective practice, however, is that it is often assessed as part of an assignment. As argued by Hobbs (2007: 413): 'Reflection and assessment are simply incompatible'. This notion is further explored by Tummons (2011), who highlights several areas for concern, including the validity of such assessments. First, can we assume that everybody involved has the same understanding of what reflective practice actually means. Fook et al. (2006) point out that reflective practice or critical reflection involves:

- a process (cognitive, emotional, experiential) of examining assumptions (of many different types and levels) embedded in actions or experience;
- a linking of these assumptions with many different origins (personal, emotional, social, cultural, historical, political);
- a review and re-evaluation of these according to relevant criteria (depending on context, purpose etc.);
- a reworking of concepts and practice based on this re-evaluation.

They continue that the ways in which reflective practice is used will vary according to the focus and purpose, and the types of processes involved. The notion of 'reflective practice', therefore, varies considerably (Fook et al. 2006). This, in its turn, has an impact on the validity of assessing reflective coursework. If the criteria for marking are somewhat loose, as we have different understandings of what is required, reliability is also affected. A further issue is that of honesty. Will students submit what they truthfully think or have experienced, or is there some chance that they write what they think the assessor may want to read, especially as the work may be graded. This type of practice could of course make a mockery of the process, with students not appreciating the purpose of reflective practice, further fuelling their resistance to participate. Despite these limitations of reflective practice (referred to by Finlay (2008) as the 'dark side'), at present it persists to be at the heart of any teacher training programme. So, what are the benefits?

The benefits of reflective practice

A significant characteristic of reflective practice is that, when used effectively, it enables the practitioner to make sense of a subject in their own way, using their own language and at a pace suitable for them at any given time. It is the actual process of making sense of a situation or idea that is important, because the process involves an individual taking responsibility for their own learning. For example, Ashby (2006: 35) considered her reflective practice, observing that 'The use of structured reflection has helped me to come to terms with situations when I was unhappy with my performance by helping me to understand what went wrong or why I behaved as I did.' The typical characteristics of this process are that she is linking her behaviour to her performance within a particular situation. Exactly what this process involved for her is not important; what is significant though is that a structured process is being undertaken. What is notable about the reflection in this example is the fact that the criterion selected for comparative purposes appears to be ipsative in nature. She has not commented that in comparison to others she did less well (norm referenced), or that she failed to meet set criteria (criterion referenced). Instead she is concerned primarily with her performance at that time and her anticipated future performance. This is a significant feature of reflective practice where an individual sets their own criteria. Arguably, a person is in the best position to understand themselves, make sense of their unique situation and to be brutally honest with themselves. Although valid, any external observations would, arguably, lack the insight an individual has of themselves. Ashby continues, 'By reflecting on problems I have been able to find a way forward, by recognising and filling the gaps in my knowledge or planning how I would deal with a similar situation in the future' (ibid.). In this case the benefits of reflective practice were that any gaps in knowledge were identified and the individual was able to consider the temporal aspect of the situation. As with the vague nature of ipsative criteria, although planning for a similar situation was possible, the exact timing, however, was not determined; instead the plan was made for the next occasion whenever that may be. The process of reflective practice can be at times vague and the path meanders as the individual muses upon a situation and returns to it from a number of possible perspectives. It is the fact that there are no hard and fast rules in relation to the thinking practice that frees the practitioner. It is not rule bound and can be an empowering experience. Reflective practice is, arguably, the most student-centred aspect of teaching and learning and, as such, is a very valuable learning tool.

Connectedness, honesty and perspectives

Reflective practice works on a principle of connectedness. McIntosh (2010: 37) argues that the 'ideas of connectedness suggest that we are always in relation with someone'. In a social event, the patterns that appear as a consequence of the connections made between people can be observed. As a practitioner recounts an experience, it is the connections that are identified. Arguably, without a connection there can be no meaningful interaction. A series of connections provide possible causes and effects; these represent a social phenomenon. The question of causation within the social event is one that occupies the mind during reflection. For example a person could ask 'What did I do that brought about that situation?', or 'Did the emergence of that situation actually have anything to do with me?', or 'Why did she behave that way to me?' Reflective practice enables a person to situate themselves within the pattern of connections and to recognise the changing shape of the patterns through time.

Involved in the process of reflective practice is the requirement of honesty. A person is capable of being dishonest with themselves, and this can manifest itself as denial at one end of the bipolar axis, through to having delusions of grandeur at the other end. However, Moon (2006: 94) suggests that there is 'not necessarily one truth about an event ...'. This represents what she refers to as a 'fundamental point of understanding': it is a principle that should be discovered by the reflective practitioner rather than it being taught. This understanding occurs as a part of the development of an individual in relation to ideas of truth and uncertainty in knowledge (ibid.). Honesty is the gateway to truth, and acknowledging the existence of a situation opens that gate. For example, returning to Ashby (2006), she notes that 'I have then been able to discuss those problems or mistakes more easily with others, rather than trying to bury and deny them'. Two things occur in this instance: first the individual's ability to be honest, and second the sharing of reflective practice with others. This represents another form of connectedness, where a person shares with others their thoughts within a social group, as opposed to considering the events alone. The difference here is that the connectedness takes place both during and after the event; the name we give to this phenomenon is Co-operative Reflective Practice. To qualify as such, the population of the group should be made up only of the participants of the previous encounter or situation. As a consequence, the sharing of ideas and listening to others provides multiple perspectives. If the group were made up of those not present at the time (with the exception of the individual who experienced that situation), then the participants would only be in a position to interpret the situation;

this we call Interpretive Reflective Practice. This idea correlates to Brookfield's 'critical lenses' (Roffey-Barentsen and Malthouse 2009). Here, four critical lenses (or perspectives) are considered in relation to reflective practice in relation to teaching:

1   the point of view of the teacher;
2   the point of view of our learners;
3   the point of view of our colleagues;
4   the point of view of theories and literature.

Essentially there are two types of reflecting, alone or with others, and each offers varying advantages. On the one hand, reflecting alone promotes cognition, where the person is forced to think through an issue without relying on the observations, insights and directions of another. The person is able to stay in their own model of the world, without the need to explain a situation in a way that another would be able to understand. Issues are not clouded by observations from others, who may have a differing view of a situation. The speed of thought is appropriate for the individual, without the need to keep up with, or slow down for, another. The person can make sense of a situation in their own way. They can understand a situation employing their associated cognitive and affective responses in relation to their life's experience. For example, if they are experiencing issues of transference, then they can think though the situation without the need to explain the myriad of emotions and thoughts to another. In other words, no explanation or justification is necessary and, as a result, the process of reflection is not constrained by others.

On the other hand, reflecting with others enables a person to view things from another's perspective. Where those who were involved in the original scenario are also involved in the group reflection, it may be possible to identify causation and reactions. Intentions can be discussed and, where misunderstanding occurred, the causations may be identified. According to Roffey-Barentsen and Malthouse (2009: 8), by referring back to Brookfield's third lens above ('The point of view of our colleagues'), 'you are able to enter into a more critical conversation about your practice'. As a consequence, areas can be highlighted that, until that time, had not been considered.

So far this chapter has considered the terms Co-operative Reflective Practice (where only those present during the original situation participate) and Interpretive Reflective Practice (where a group is made up of those not present at the time, with the exception of the person who experienced the situation). A combination of these groups we refer to as Collaborative Reflective Practice. Here, the group is comprised of some who experienced the situation and some who have not. The benefit of Collaborative Reflective Practice is that the dynamics can be more varied, where, for example, those who experienced a situation could perhaps interpret it in various ways. Equally, those who were not present during the original situation are able to draw on and consider a wide selection of accounts. Collaborative Reflective Practice offers a comparatively wide variation of interpretations, explanations and analysis. Last, we further offer the term Representative Reflective Practice for a form of reflection by proxy. This is used to identify reflective practice in relation to a situation or experience not originally witnessed by the reflective practitioner at that time, but which is experienced after the original event. For example, the participants view a clip from the Internet, a film or read a chapter from a book; they then discuss how they felt about the characteristics of the experience. The advantage of Representative Reflective Practice is that it can create an experience in a safe environment and focus on a topic that is unlikely to be experienced by any member of the group. This may lead to a rich learning experience that may otherwise have not been possible. So to recap, the forms of reflective practice offered are:

- *Co-operative Reflective Practice* – Group reflective practice where all participants were involved.
- *Interpretive Reflective Practice* – Group reflective practice where only one individual experienced the situation and the remainder of the participants interpret the account.
- *Collaborative Reflective Practice* – Group reflective practice where there is a mix of participants, some involved with the original event and some not.
- *Representative Reflective Practice* – Group reflective practice by proxy.

How does this relate to science teachers?

As we have recognised, science teachers may find it challenging to engage with reflective practice. As a trainee teacher, it can be a lonely experience, especially if it is not clear exactly where to start and in which direction to go. Regular meetings with peers or colleagues may provide a suitable launch pad from which to embark on the reflective journey. It is as a result of working with others, sharing ideas and discussing the processes that enables an individual to familiarise themselves with the notion of reflective practice. Expecting a student (scientist or other) to effectively engage with reflective practice is unrealistic. Group reflective practice offers an environment in which an individual can expect suitable support from others.

Situated Reflective Practice

We offer a recently-developed form of reflective practice referred to as Situated Reflective Practice (SRP) (Malthouse 2012). SRP introduces a variation in relation to the traditional forms of reflective practice, because it considers the way in which a social group or an organisation is acting, and the influence this has upon an individual; as opposed to considerations of the direct actions of an individual. Within SRP, the focus of attention may relate to a phenomenon that has very little, or indeed nothing, to do with the individual experiencing it. In fact, a person may have no control in relation to the situation about which they have been made aware. The situation which has now been recognised by the individual may not have the potential to influence them at that time, but typically it may. There are five characteristics of SRP, described by Malthouse (2012: 298) as:

1   *Passive observation of an action* – The individual may take a casual interest in the situation. Significantly, the individual has little of no control of the phenomenon; this is independent of the person observing it.
2   *Experienced by proxy* – The individual may consider that the situation has the potential to influence them, albeit at a later stage.
3   *Found in social/organisational situations* – The phenomenon represents a trend or change, for example the restructuring of an organisation. It has the possibility to affect a large number of people.
4   *Can occur at the time or after* – The temporal nature is relatively slow, as it represents a social or organisational change. The reflections would typically take place as they occur.
5   *Locus of power separate from the person experiencing it* – The individual has no say or influence in the situation whatsoever. As a result, the individual can either choose to take action or wait as the situation develops.

The process of Situated Reflective Practice is represented by five stages shown in Figure 12.1.

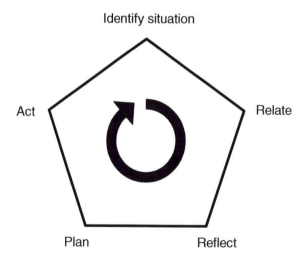

Identify situation

Act

Relate

Plan

Reflect

Figure 12.1 Situated Reflective Practice model.

Source: Malthouse (2012: 300).

Stage 1   *Identify Action* – The individual identifies an 'emerging phenomenon'. This can be a social or organisational situation, but could be anything that the individual has become aware of.

Stage 2   *Relate* – Next the individual makes the link between their situation and the emerging phenomenon. At this stage there is a realisation that the emerging situation could potentially affect them.

Stage 3   *Reflect* – The individual considers the situation from various perspectives.

Stage 4   *Plan* – A plan is decided upon which is dependent upon Stage 3 above.

Stage 5   *Act* – The individual now acts as appropriate, based upon the above plan.

The benefit of Situated Reflective Practice is that it offers a different perspective to an individual. Rather than restricting considerations within the classroom or teaching practice, SRP enables an individual to consider the larger picture, to identify and analyse a given situation and to prepare for it. It allows a person to regain autonomy in an environment in which they may have very little or no control.

Conclusion: reflective practice at the hub

In essence, reflective practice supports learning and teaching, and, without it, practitioners' teaching practice would be poorer. Kane *et al.* (2004: 283) offer a model that situated reflective practice at the hub of five 'interrelated dimensions', namely: subject knowledge, skill, interpersonal relationships, teaching/research nexus and personality, as shown in Figure 12.2.

Reflective practice is understood by Kane *et al.* (2004: 283) to support the interrelated dimensions listed above. They observe that 'Reflection lies at the hub of our model and we propose that it is the process through which our participants integrate the various dimensions'. Reflective practice is highly regarded as a means by which practitioners '… used differing types of reflection to improve their understanding of dimensions in their teaching'. It had the effect of increasing the understanding and relationships of the above dimensions so that teachers were better situated to understand the relationships between each. Reflection has been placed at the hub of this model because of its central role in ensuring that '… reflective, self-critical practice … enables our participants to understand and reconcile the various dimensions of teaching and to establish excellence' (ibid.).

Key questions

1   What support do you think trainee science teachers require to embrace an alternative paradigm to that of positivism?

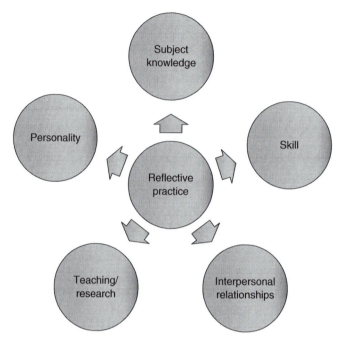

Figure 12.2 Five interrelated dimensions.

Source: Kane *et al.* (2004: 283).

2   How can reflective practice be best assessed when it forms an integral part of an Initial Teacher Training programme?
3   How can science teachers be encouraged to continue to engage in reflective practice following completion of their Initial Teacher Training?

Further reading

McGregor, D., and Cartwright, L. (2011) *Developing Reflective Practice: A guide for beginning teachers.* Maidenhead: Open University Press. This publication provides an informative mix of theoretical frameworks and practical scaffolds, explained through examples of school-based practice.

Moon, J. A. (2004) *A Handbook of Reflective and Experiential Learning: Theory and practice.* Abingdon, Oxon: Routledge-Falmer. This text offers an academic approach to theories and models of reflective practice. Written for those who wish to explore the subject to further develop their practice.

Roffey-Barentsen, J., and Malthouse, R. (2012) *Reflective Practice in Education and Training.* London: Sage. A step-by-step introduction to the theory and application of reflective practice. Written in an accessible language, this publication guides those new to the subject and offers a new perspective on the topic.

References

Argyris, C., and Schön, D. (1992) *Theory in Practice: Increasing professional effectiveness*. San Francisco, CA: Jossey-Bass.

Ashby, C. (2006) The benefits of reflective practice. *Practice Nurse*, 32(9): 35–7.

Boud, D., Keogh, R., and Walker, D. (1985) Promoting reflection in learning: A model. In D. Boud, R. Keogh and D. Walker (eds) *Reflection: Turning experience into learning*. London: Kogan Page.

Brookfield. S. D. (1987) *Developing Critical Thinkers: Challenging adults to explore alternative ways of thinking and acting*. Buckingham: Open University Press.

Cohen, L., Manion, L., and Morrison, K. (2000) *Research Methods in Education* (5th edn). London: Routledge Falmer.

Ecclestone, K. (1996) The reflective practitioner: Mantra or model or emancipation? *Studies in the Education of Adults*, 28(2): 146–61.

Edwards, G., and Thomas, G. (2010) Can reflective practice be taught? *Educational Studies*, 36(4): 403–14.

Efe, R. (2009) Science student teachers' attitudes towards reflective practice: Differences in subjects and grades. *Cypriot Journal of Educational Sciences*, 4: 72–86.

Finlay, L. (2008) *Reflecting on 'Reflective practice'*. Available online at: www.open. ac.uk (accessed 11 November 2012).

Fook, J., White, S., and Gardner, F. (2006) Critical reflection: A review of contemporary literature and understandings. In S. White, J. Fook and F. Gardner (eds) *Critical reflection in health and social care*. Maidenhead: Open University Press.

Gault, A. (2011) How does your teacher identity fit with the culture of teaching and the organisation? In D. McGregor and L. Cartwright (eds) *Developing Reflective Practice: A guide for beginning teachers*. Maidenhead: Open University Press.

Gibbs, G (1988) *Learning by Doing: A guide to teaching and learning methods*. Oxford: Further Education Unit.

Hobbs, V. (2007) Faking it or hating it: Can reflective practice be forced? *Reflective Practice*, 8(3): 405–17.

Hyatt, D. F., and Beigy, A. (1999) Making the most of the unknown language experience: Pathways for reflective teacher development. *Journal of Education for Teaching*, 25(1): 31–40.

Kane, R., Sandretto, S., and Heath, C. (2004) An investigation into excellent tertiary teaching: Emphasising reflective practice. *Higher Education*, 47: 283–310.

Kemmis, S. (1985) Action research and the politics of reflection. In D. Boud, R. Keogh and D. Walker (eds) *Reflection Turning Experience into Learning*. London: Kogan Page (pp. 139–64).

Kind, V., and Taber, K. (2005) *Science. Teaching School Subjects 11–19*. London: Routledge.

Kinsella, E. A. (2007) Embodied reflection and the epistemology of reflective practice. *Journal of Philosophy of Education*, 41(3): 395–409.

Kolb, D. A. (1984) *Experiential Learning: Experience as the source of learning and development*. New Jersey: Prentice Hall.

Larrivee, B. (2000) Transforming teaching practice: Becoming the critically reflective teacher. *Reflective Practice*, 1(3): 293–307.

Loughran, J. J. (2007) Science teacher as learner. In S. K. Abell and N. G. Lederman (eds) *Handbook of Research on Science Education*. New Jersey: Lawrence Erlbaum.

McIntosh, P. (2010) Action research and reflective practice. In *Creative and Visual Methods to Facilitate Reflection and Learning*. Abingdon, Oxon: Routledge.

McIntyre, D. (1993) Theory, theorising and reflection. In J. Calderhead and P. Gates (eds) *Initial Teacher Education in Conceptualising Reflection in Teacher Development*. London: Falmer Press.

Malthouse, R. (2012) *Reflecting Blues: Perceptions of policing students with regard to reflective practice and associated skills*. Saarbrücken: Lambert Academic Publishing.

Moon, J. A. (2006) *Learning Journals: A handbook for reflective practice and professional development*. Abingdon, Oxon: Routledge.

Olson, J. K., and Finson, K. D. (2009) Developmental perspectives on reflective practices of elementary science education students. *Journal of Elementary Science Education*, 21(4): 43–52.

Parsons, M., and Stephenson, M. (2005) Developing reflective practice in student teachers: collaboration and critical partnership. *Teachers and Teaching: Theory and Practice*, 11(1): 95–116.

Paul, A. (2010) Mentoring reflective practice in pre-service teachers: A reconstruction through the voices of Australian science teachers. *Journal of College Teaching and Learning*, 7(9): 1–18.

Quinn, F. M. (2000) Reflection and reflective practice. In C. Davies, L. Finlay and A. Bullman (eds) *Changing Practice in Health and Social Care*. London: Sage.

Roffey-Barentsen, J., and Malthouse, R. (2009) *Reflective Practice in the Lifelong Learning Sector*. Exeter: Learning Matters.

Schön, D. (1987) *Educating the Reflective Practitioner*. San Francisco, CA: Jossey-Bass.

Tummons, J. (2011) 'It sort of feels uncomfortable': Problematising the assessment of reflective practice. *Studies in Higher Education*, 36(4): 471–83.

Zukas, M., Bradbury, H., Frost, N., and Kilminster, S. (2010) Conclusions. In H. Bradbury, N. Frost, S. Kilminster and M. Zukas (eds) *Beyond Reflective Practice*. London: Routledge.

Chapter 13

School science and technology

David Barlex and Marion Rutland

Introduction

In this chapter we will consider the possibility that pupils can use the learning that is taking place in the science curriculum (or has taken place) to enhance their learning in design & technology. The chapter begins with a brief history of the relationship in schools between science and 'design & technology'. We then consider approaches to enhance the relationship. The bulk of the chapter presents examples of design & technology activities that can be informed through links with science. Finally we consider the feasibility of such an approach.

The science–design & technology relationship

In the world outside school there is a strong relationship between science and technology. In considering the nature of technology and the way it evolves, Brian Arthur (Arthur 2009) argues that technology can be seen as the exploitation of phenomena revealed by science. He rejects a simplistic 'technology is applied science' view, but is adamant that it is from the discovery and understanding of phenomena that technologies spring. He notes that:

> It should be clear that technologies cannot exist without phenomena. But the reverse is not true. Phenomena purely in themselves have nothing to do with technology. They simply exist in our world (the physical ones at least) and we have no control over their form and existence. All we can do is use them where usable. Had our species been born into a universe with different phenomena we would have developed different technologies. And had we uncovered phenomena over historical times in a different sequence, we would have developed different technologies.
>
> (Arthur 2009: 66)

Arthur (ibid.) also discusses the almost bipolar relationship that humans appear to have with technology, in terms of our trust for the natural compared to our suspicion of the artificial.

These two views, that technology is a thing directing our lives, and simultaneously a thing blessedly serving our lives are simultaneously valid. But together they cause unease, an ongoing tension, that plays out in our attitudes to technology and in the politics that surrounds it.

(Ibid.: 214)

... we trust nature, not technology. And yet we look to technology to take care of our future – we hope in technology. So we hope in something we do not quite trust ...

(Ibid.: 215)

If we follow Arthur's line of reasoning – and he does make a persuasive case – we must ask how is this intimate, if somewhat fraught, relationship reflected in the school curriculum? The picture that emerges with regard to the reality of practice is very different. The Interaction Report (Barlex and Pitt 2000) noted 'In schools a separate and almost unrelated relationship exists between science and design & technology in direct contrast to the that between science and technology in the world outside school' (ibid.: 5). The report found that 'Each group [science and design & technology teachers] holds coherent views about its own subject, but a variety of views about the other subject few of which coincide with the view from inside that other group' (ibid.: 24). To some extent both subject groups share the blame for this mutual ignorance but, given that a science publication like *New Scientist* always carries a technology section and often deals with matters technological, one could argue that the science community should take more interest in and be more informed about what's happening in design & technology lessons.

The report *Becoming an Engineering College* (Barlex 2005) explored the collaboration that was taking place in schools with the specific brief to develop cross-curricular links. Teachers cited the prescriptive nature of the National Curriculum in mathematics and science as a barrier to developing links with design & technology. The lack of quality time to develop and plan collaborative activities was also cited as a reason for a lack of cross-curricular work. The nature of some of the difficulties encountered when science and design & technology teachers attempt to collaborate has been revealed through a case study approach (Lewis *et al.* 2007). This research showed that the misaligned views of the subjects collaborating led to a situation where the teachers challenged each other's subject knowledge and pedagogical culture. For example, whilst food technology teachers were content to discuss the nutritional value of food stuffs at the macro level of sugars and starches, the science teachers insisted that it was necessary to look at the behaviour of foods in digestion at the molecular level. While these two approaches could have been seen as complementary and building a more complete picture, in this situation the difference of approach led to professional antagonism. Similarly, within electronics, while the design & technology teachers argued that pupils only needed to understand functional characteristics of components to be successful in simple circuit design, the science teachers thought it necessary for pupils to understand why a component behaved as it did. Again these positions can be seen as complementary and potentially mutually reinforcing, but these different positions led to disagreement over learning outcomes. Underlying both sets of disagreements are misunderstandings of the nature and purpose of knowledge within the two curriculum areas.

Science is primarily concerned with exploration and explanation of what exists, developing and using declarative knowledge, whereas design & technology is concerned with the conception of what yet does not exist and how it might be brought into existence, requiring and developing normative knowledge. Marc De Vries (2007), a well-known and authoritative philosopher of technology explains this through the statement: 'There is no such thing as a good electron!' An electron just is and behaves as electrons do. But, of course, once we decide to utilise the phenomenon of electron behaviour we are in different territory. In providing the energy to make generators rotate and produce electric current, we have to make judgments about energy sources that can be used and their impacts. The way the electric current is produced can, according to your perspective, actually be good or bad. And we also have to make judgments about the uses to which the electricity generated might be put: at opposite extremes it can be used to end life through the use of the electric chair for capital punishment, or save life by operating a premature baby incubator.

More recently, the STEM Pathfinder project reported more success in developing a useful relationship between the STEM subjects (science, technology, engineering and mathematics), including that between science and design & technology.[1] However even in schools committed to developing such relationships there were difficulties. As the evaluation reported:

> The major challenges faced by teachers were finding time to meet together and plan activities, timetabling activities, and getting other staff involved in the activities. Schools used different approaches to overcome these challenges, including: finding time by meeting after school, using STEM training days to plan, and creating a funded STEM post to coordinate planning; using cross-curricular days and delivering activities in the summer term to overcome timetabling issues; and targeting specific staff or organising joint STEM CPD to get other colleagues involved.
>
> (NFER 2009: v)

So the track record for pupils using science learning in their design & technology lessons is not good. We move on to consider two views that give insight into how the situation can be improved.

Enhancing the design & technology–science relationship

Janet Ainley and colleagues (Ainley *et al.* 2006) have considered the problem of cross-curricular relationships by conceptualising it as a planning paradox. If teachers plan from tightly-focused learning objectives, the tasks they set are likely to be unrewarding for pupils and (in the cases they are considering) mathematically impoverished. If teaching is planned around engaging tasks, the pupils' activity may be far richer, but it is less likely to be focused, and learning may be difficult to identify or assess. They suggest that the two constructs of purpose and utility offer a framework for task design that may resolve the planning paradox. They argue that it is possible to engage the utility of some subjects in pursuing the learning purposes of another subject.

Hence it should be possible to capitalise on the utility of, for example, science in pursuing the learning purposes of design & technology. A fundamental purpose of design & technology is for pupils to learn how to make design decisions so it is worth elaborating the sorts of design decisions that pupils are expected to make. The decision-making that pupils have to undertake when they are designing and making has been described as involving five key areas of interdependent design decision (Barlex 2004, 2007): conceptual (overall purpose of the design, the sort of product that it will be); technical (how the design will work); aesthetic (what the design will look like); constructional (how the design will be put together); and marketing (who the design is for, where it will be used, how it will be sold). This approach can be represented visually as a pentagon diagram as shown in Figure 13.1.

This inter-dependence of the areas is an important feature of design decisions; hence the lines connect each vertex of the pentagon to all the other vertices. A change of decision within one area will affect some if not all of the

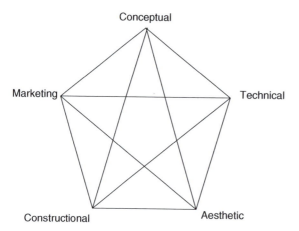

Figure 13.1 The design decision pentagon.

design decisions that are made within the others. The teacher usually identifies the sort of product the pupils will be designing and making. Hence it very difficult for pupils to engage in conceptual design, but there are still many opportunities for making design decisions in the other areas. It is the juggling of these various decisions to arrive at a coherent design proposal that can then be realised to the point of a fully-working prototype that provides the act of designing and making with such intellectual rigour and educational worth and an essential part of design & technology education. It is of course necessary to ensure that the outcomes of designing and making meet criteria that establish their viability and worth. The following is offered as an initial list for further discussion.

- Embedded within them must be high learning value.
- The resources required must be such that the outcomes are affordable and the processes by which they are achieved manageable.
- It is important that they are intriguing, desirable and non-trivial as far as pupils are concerned. Hence the outcomes should have usefulness whatever the focus area of learning.
- To relate this designing and making to the world outside school, it will be important for pupils to consider life cycle analysis of similar manufactured items.

It is important that such decisions are genuine design decisions and not simply technical decisions contrived to support learning in science. Ainley and her colleagues (Ainley *et al.* 2006) also argue that there is mutual benefit in this arrangement. In utilising science, pupils will become more adept at this subject while at the same time enhancing their ability in design & technology. This is an important point. If teachers of different subjects are to develop cross-curricular relationships, then it is essential that all the subjects involved benefit.

Torben Steeg, widely regarded as a national expert in the teaching of electronics, systems and control, and modern manufacturing, also has a strong background in science education, having spent the early part of his teaching career as a physics teacher. He has clear views on the way science learning can inform pupils' learning in design & technology (Banks and Barlex 2013). Paramount he thinks is establishing an understanding of scientific method, or, even better, inculcating scientific thinking, i.e. the ability to approach a question in design & technology with a desire for empirical evidence; the attitude of 'let's find out'. Torben notes that it will be important for pupils to understand when a scientific approach is likely to be useful and when it is not. Exploring the properties of a range of materials to identify which will have the required strength and stiffness is an obvious example of a worthwhile empirical investigation, particularly if the pupils are just getting to grips with the difference between strength and stiffness. Developing the overall appearance of a product such that it is likely to appeal to a particular user group is much less likely to rely on a scientific approach.

The next section will describe examples of design & technology activity in which pupils will be able to use their science knowledge and understanding to the advantage of learning in both subjects. The activities will require pupils to make design decisions and will be scrutinised from three perspectives. First, that provided by Brian Arthur, in that the activity should deal to some extent with the exploitation of a scientific phenomenon. Second, that there is a utility purpose relationship, as envisaged by Janet Ainley and colleagues, in the way science knowledge and understanding is used to pursue a purposeful design & technology activity. Third, as proposed by Torben Steeg, that the activity embraces scientific thinking as well as scientific knowledge.

Examples where science informs design & technology

Example 1: Designing and making a heating/cooling device

In this example we start by invoking the exploitation of phenomena with regard to the Peltier Effect. This effect is enshrined in a solid state device that, when activated, transfers heat from one side of the device to the other side against the temperature gradient. This results in a 'cold on one side hot on the other side' sensation which is likely to intrigue 14-year-olds. Note that this phenomenon is outside their usual science curriculum, hence it provides an opportunity for an investigation in which the results are not already known, as is the case in most standard science investigations. Understanding the physics of the Peltier device in terms of a scientific explanation is *not* required. Hence, although the underlying science will not be taught until pupils are several years older, it is possible for Year 9 pupils to investigate and identify the performance characteristics of the device, and that is the knowledge needed to be able to take action in terms of making design decisions. Pupils are required to investigate the extent to which the phenomenon can be exploited: What arrangements are required to get the cold side really cold and the hot side really hot, i.e. maximise the effect? This approach will not only call upon pupils' understanding of heat and temperature, but also on their ability to think scientifically in devising experiments to probe the performance of the Peltier device.

Philip Holton has used this approach with Year 9 pupils. In his classroom, pupils have designed and made a variety of cooling devices for different purposes that they consider worthwhile, including cooling drinks and maintaining an organ for transplant at the correct temperature during transportation. This activity will meet Ainley's purpose – utility conditions, as the pupils will be using the science knowledge and skills to inform their technical design decisions as they devise a device of their own choosing. Without this science learning it is unlikely that the pupils would be able to produce feasible functioning designs that they could then make. We must ask: Where is the advantage is for the science teacher? The science teacher can take the opportunity to question pupils about the way the devices they have designed work.

It is likely that the pupils will be eager to explain. Their answers will reveal the extent to which pupils understand the difference between heat and temperature, and the mechanisms of heat transfer. In some cases, the pupils will have been able to consolidate and advance their understanding of energy, and this will be revealed by their answers. In other cases they will reveal conceptual confusion, which the question answer session may go some way to address – or at least inform both the science teacher and the pupil that there is still some way to go in developing a sound understanding.

Example 2: Lighting

A variety of phenomena are exploited in the production of lighting devices. Candles and kerosene lamps exploit the combustion of hydrocarbons, although, in addition to producing light, the devices that use this phenomenon produce toxic fumes and are fire hazards. Parts of the world which have no mains electricity still use such devices. The gas mantle exploited the combustion of a hydrocarbon (coal gas) to heat a fine mesh of thorium and cerium oxide which glows brightly producing white light. This form of heating was used extensively for street lighting in England until the early 1900s, until the widespread introduction of mains electricity. It was still common in some homes in England until the 1960s. Until quite recently, filament lamps powered by mains electricity were used to provide lighting in most homes in England. The passage of the electric current through the thin filament caused it to become hot and glow brightly. But now these are giving way to lower energy consumption fluorescent bulbs and light emitting diode arrays which do not utilise the heating effect of the electric current.

Such a brief history of phenomenon exploitation in providing lighting can form the background to a design & technology task in which pupils are required to design and make simple lighting devices. A comparison between filament lamps and light emitting diodes (LEDs) provides a practical entry point to this narrative about the way the provision of domestic lighting has and is undergoing change. The way in which filament lamps work is relatively easy for pupils to understand, and it is not difficult for them to appreciate how inefficient such lamps are, in that only a fraction of the energy consumed is used in providing light. Most of the energy is used in bringing the filament up to the temperature at which the filament begins to glow. And of course pupils can feel filament lamps becoming hot. LEDs on the other hand do not rely on a heating effect to produce light.

So if pupils are designing and making an LED-based light for, say, task or mood lighting, it is possible to support this with science-based investigations into filament lamps and LEDs. This is likely to be of interest to the physics teacher who might even engage pupils in a comparison of filament lamps and LEDs as part of the pupils' physics course. This comparison could be extended into the design & technology lessons to include which sorts of LEDs give the most light, how many LEDs could be powered from a 9V battery, the protection they might require, and the pros and cons of series versus parallel arrangements. This approach will not only call upon pupils' understanding of electricity, but also on their ability to think scientifically in devising experiments to probe the performance of different circuitry. And clearly the approach involves pupils using the utility of science in being able to pursue the purpose of designing and making an LED lamp.

Note that, in the approach described above, the initial investigation starts in physics and then moves to design & technology. Questioning the pupils about the functioning of their finished lamps back in physics would reveal the extent to which the pupils really understood some basic ideas about electricity. But this activity need not stop at physics. Recent work by Philips has explored the use of bioluminescence in domestic lighting.[2] Here there is the opportunity for design & technology to link with the emerging field of biomimetics –adopting and adapting biological systems for use in technologies. Pupils could compare the LED-based lights that they had produced with the lights being developed by Philips which use light-producing bacteria and need no wires, batteries or connection to an electricity grid. The energy comes from the bacteria's food source, which the researchers at Philips suggest could come from the sludge from a methane digester.

Example 3: Baking bread

Many cultures have developed the technologies needed to bake bread. In some of these cultures bread is a staple food. There is variation according to locality, and bread has cultural significance for many communities. The scientific phenomena underlying the production of bread has been known for many years and is not particularly complex. It involves understanding the behaviour of raising agents on flour-water mixtures, and the action of heat on such mixtures which give rise to foams with a hard crust. Some breads use only the steam produced by the heating action to create the foam, while others use yeast to generate carbon dioxide gas which contributes to the gas in the foam before baking. A variety other ingredients (e.g. seasonings, herbs, nuts, olives, dried fruit) can be added to the mixture to give added texture and flavour.

Hence it is not difficult for pupils to devise experiments which enable them to investigate the role of various ingredients and processes, and the influence they have on the nature of the final product. Solid foams are an important structural form in many food products: breads, cakes, pastries, biscuits. The texture will depend on the structure of the foam in terms of the size of the bubbles in the gas phase and the elasticity of the material making up the solid phase. The temperature and time of cooking will affect these.

Such an approach to the properties of food requires pupils to understand the idea of properties, and that the overall properties of complex mixtures will depend on both the structure of the mixture and the properties of the materials within the mixture. This is made even more complex by the fact that the baking process can affect both the properties and the structure. Unpacking why a particular recipe gives the results it does is not a trivial task and requires science understanding. Hence such an approach within food technology would meet Arthur's view that technology exploits science phenomena. The proof of any pudding is in the eating, and pupils will learn about a range of food tasting techniques, e.g. ranking, preference and difference tests. If pupils adopt a scientific view of bread and consider their own efforts to devise bread products as an exercise in 'solid foam design', then clearly the utility is science is playing out in pursuing a design & technology purpose. This approach may be new to some food technology teachers, but it has the potential to strengthen links between science and design & technology, not only in terms of requiring pupils to use science knowledge and understanding, but also in providing science teachers with the opportunity to discuss the design activity as a formative assessment exercise, probing pupils' understanding of chemical reaction, properties of gases and enzyme behaviour.

Example 4: Protective textiles

An obvious area of protection that pupils might consider is body armour as used by police forces and the military. A modern material used for this purpose is Kevlar – a synthetic polymer that can be spun to form a fibre that has a high tensile strength-to-weight ratio, being five times stronger than steel on an equal weight basis. It can be woven into ropes and fabric sheets that are used in composite materials.

Pupils are likely to have little experience of this material, although visitors from the police or armed forces may be able to provide examples which pupils can handle. Pupils are much more likely to have experience of everyday materials that were used prior to the advent of Kevlar. Handling and investigating such materials is an important precursor to considering the properties of Kevlar if pupils are to appreciate both the qualitative and quantitative differences between these materials and Kevlar. Kevlar is stronger. How much stronger? Kevlar is lighter. How much lighter? Kevlar is stiffer. How much stiffer? These comparisons require both scientific knowledge and understanding of property concepts, and the ability to think scientifically in ensuring that the comparisons are valid. Thus far the pupils are appreciating the remarkable properties of Kevlar; now is the opportunity for them to consider what they would design if they had access to the material.

Moving from law enforcement and military situations into sporting activities would provide contexts where the protection would be required to prevent damage from bats, balls, falls and collisions with other players. Such protective items are often required to have visual appeal as well as functional effectiveness, so there are opportunities for pupils to exhibit both visual flair as well as science knowledge and understanding. It is of course important that the aesthetic decisions made by the pupils are not in conflict with the performance characteristics of Kevlar. Here pupils are engaged with the utility of science as they develop and justify technical design decisions. The work can be extended by asking the pupils about changes they might make to their design proposals if they were to use different materials. An interesting material to consider is spider silk – five times stronger than steel, tougher than Kevlar and highly elastic, so potentially extremely useful if only it could be manufactured. Such speculations will again assess their appreciation of the scientific knowledge and understanding. If it is important that pupils design *and* make protective textile items, then protection from the cold provides the opportunity for pupils to test a variety of fabrics and fabric combinations for their ability to insulate, and then go on to choose particular materials or combinations for their designs, based on the results of their investigations. Here we have a combination of science knowledge and understanding with regard to mechanisms of heat transfer, and the ability to carry out scientific investigations providing the utility needed to pursue the design & technology purpose. Although important, thermal properties are not the only considerations. Flexibility will need to be considered if the garments produced have to fit over moving parts, and some degree of handling capacity is required as is the case for gloves. And in some cases water resistance or waterproofing will be required as well. So several phenomenon will be under exploitation here.

Conclusion: reality or fantasy?

So, as we near the end of the chapter, we must ask to how feasible are the approaches suggested above? Their aim is to forge a robust and mutually-reinforcing relationship between science and design & technology. If a science department by operating in a 'subject silo' can perform well and meet the metrics required by the school leadership, then what is the incentive to change? Why bother to talk with design & technology teachers and understand their curriculum area and what they are trying to achieve? It will take both time and effort – and any return on this investment may be uncertain, and it will probably take some time to yield any measurable result.

We argue that by *not* taking this opportunity, science teachers are in denial of a highly-significant vested self-interest. Supporting the use of science learning within design & technology provides pupils with the opportunity to consolidate their scientific knowledge and understanding and, in some cases, reveal misunderstandings that are unlikely to be divulged in science lessons. Revealing the usefulness of science in this personal and immediate way is likely to convince pupils that scientific knowledge and understanding is worthwhile in its own right. Hence this will provide motivation not only to learn science in the short term, but also to consider it as a worthwhile future qualification that might lead to a science-based career. This is an important feature in justifying the making of links between science and design & technology. The ROSE (Relevance of

Science Education) Project (Jenkins and Pell 2006) into young people's attitude has revealed that, while many young people consider science important and useful for the wellbeing of society, they are not predisposed to study it themselves or consider a science based occupation.[3] A further point made by Robin Millar in his presidential address to the Association for Science Education in 2012 (Millar 2012) was the importance of science for all, as opposed to those pupils who will make up the next generation of scientists. We consider engaging pupils with science through design & technology can be seen as an important aspect of science for all.

It is at this point one of the authors must confess to considerable bias as, not only did he find out about design & technology when he became head of a science and design faculty, but deliberately organised the time table so that he and science colleagues could work with design & technology in teaching design and technology. It was hard work on both sides, but the impact was significant. Using science as part of the repertoire of design & technology became a natural feature of lessons, and the revelations of pupils' understanding were remarkable. Discussions around 'Will this work, why won't it work, why does it work?' gave both pupils and teachers real insight into science understanding. Ever since, he has been an advocate for developing strong links between the two subjects. This may be seen as an extreme response to the situation and one that is not easily transferrable to other situations.

So the question is: What, realistically, might science teachers do in response? Setting up regular communication such that there is a high level of mutual awareness and understanding of each other's curriculum is a definite possibility. In using this awareness and understanding to capitalise on making links between the subjects, the following strategies can be employed: talking with pupils about what they are doing in design & technology when it relates to science teaching; observing pupils in design & technology classes to see how they are (or are not) using science to support their learning; and asking pupils to bring artefacts they have designed and made into science lessons as a focus for discussing science concepts. And co-teaching on occasions with design & technology colleagues should be seen as a highly desirable possibility. We believe that such activities will pay big dividends and make both science and design & technology teaching not only more effective but also much more enjoyable.

Key questions

So having read this, what about the following issues:

1   Will you, as a science teacher, want to form a curriculum relationship with design & technology in your school? And if you do, how will you go about it?
2   Will you, as a science teacher, want to see design & technology develop so that it requires pupils to use their science knowledge and understanding? And if you do, how might you work towards achieving this?

3   If you, as a science teacher, can see advantages in forming a curriculum rela-
     tionship with design & technology in your school, how would you convince
     the senior leadership team that this is worthwhile.

Further reading

Barlex, D. (ed.) (2013) *Design and technology for the next generation*. Shropshire,
    England: CliffeCo. This is a collection of provocative pieces, written by experts in
    their field, to stimulate reflection and curriculum innovation. The book is useful
    because it has been written to help bring the teaching and educational research
    communities for design & technology closer together.
Banks, F., and Barlex, D. (2013) *Teaching STEM in secondary schools: Helping teachers
    meet the challenge*. London: Routledge. This book is relevant because it considers
    STEM from a variety of perspectives and includes many examples of links between
    the STEM subjects.
Owen-Jackson, G. (ed.) (2013) *Debates in design and technology education*. London:
    Routledge. This book is useful because it considers issues confronting design &
    technology in England as the subject faces an uncertain future.

Notes

1 Details of the STEM Pathfinder Programme and its evaluation can be found at this
   url: http://www.nationalstemcentre.org.uk/elibrary/collection/304/stem-path-
   finder-programme (accessed 20 November 2012).
2 Philips Microbial Home Project. The Microbial Home is a proposal for an integrated cyclical ecosystem where each function's output is another's input. In this
   project the home has been viewed as a biological machine to filter, process and
   recycle what we conventionally think of as waste – sewage, effluent, garbage, waste
   water. Information is available at this url: http://www.design.philips.com/about/
   design/designportfolio/design_futures/design_probes/projects/microbial_
   home/index.page (accessed 8 January 2012).
3 Information about the ROSE Project can be found at this website: http://rosepro-
   ject.no/network/countries/norway/eng/nor-Sjoberg-Schreiner-overview-2010.
   pdf (accessed 25 June 2013).

References

Ainley, J., Pratt, D., and Hansen, A. (2006) Connecting engagement and focus in
    pedagogic task design. *British Educational Research Journal*, 32(1): 23–38.
Arthur, W. B. (2009) *The nature of technology*. London: Allen Lane.
Banks, F., and Barlex, D. (2013) *Teaching STEM in secondary schools: Helping teachers
    meet the challenge*. London: Routledge.
Barlex, D. (2004) *Design decisions in Nuffield Design and Technology: Pupils'
    decision making in technology*. Research, Curriculum Development and
    Assessment Proceedings, PATT-14 Conference, Albuquerque, New Mexico,
    18–19 March.
Barlex, D. (2005) *Becoming an engineering college: A report describing emerging and
    developing good practice*. London: Specialist Schools Trust.

Barlex, D. (2007) Assessing capability in design and technology: The case for a minimally invasive approach. *Design and Technology Education: An International Journal*, 12(2): 9–56.

Barlex, D., and Pitt, J. (2000) *Interaction: The relationship between science and design and technology in the secondary school curriculum*. London: Engineering Council.

De Vries, M. (2007) Philosophical reflections of the nature of design and technology. In D. Barlex (ed.) *Design and Technology for the Next Generation*. Shropshire, England: CliffeCo.

Jenkins, E. W., and Pell, R. G. (2006) *The Relevance of Science Education Project (ROSE) in England: A summary of findings*. Leeds: Centre for Studies in Science and Mathematics Education, University of Leeds.

Lewis, T., Barlex, D., and Chapman, C. (2007) Investigating interaction between science and design and technology (D&T) in the secondary school: A case study approach. *Research in Science and Technological Education*, 25(1): 37–58.

Millar, R. (2012) Association for Science Education Presidential Address 2012: Rethinking science education: Meeting the challenge of 'science for all'. *School Science Review*, 93(345): 21 30.

NFER (2009) *Evaluation of the 2008–09 DCSF-funded Specialist Schools and Academies Trust STEM Pathfinder Programme*. Compiled by Iain Springate, Jennie Harland, Pippa Lord and Suzanne Straw. Slough: NFER.

Chapter 14

Sex education and science

Neil Taylor, Frances Quinn and Judith Miller

Introduction

Our starting point for this chapter is that sex education should have as its goal the sexual health, in its broadest sense, of all students, regardless of sexual orientation. We adopt the holistic view of sexual health as defined by the World Health Organization:

> Sexual health is a state of physical, emotional, mental and social well-being in relation to sexuality; it is not merely the absence of disease, dysfunction or infirmity. Sexual health requires a positive and respectful approach to sexuality and sexual relationships, as well as the possibility of having pleasurable and safe sexual experiences, free of coercion, discrimination and violence. For sexual health to be attained and maintained, the sexual rights of all persons must be respected, protected and fulfilled.
>
> (World Health Organization 2002)

In this chapter, we explore through the literature some of the contemporary issues associated with this kind of sex education. We focus in particular on the United Kingdom and Australia as two comparable multicultural nations with secular education systems facing similar issues in regard to sex education.

Sex education in schools: we need it

The needs and benefits for young people to be knowledgeable about their growing and changing bodies are indisputable (Bearinger *et al.* 2007; Jones and Hillier 2012; Goldman 2012). As outlined by Goldman (2008), young people are maturing earlier and modern society is commodifying sex, but many parents do not teach their children about sexuality. Hence formal schooling has a large part to play.

For example, in a study of 3,334 13–17 year olds from ten English urban and suburban secondary schools, Newby *et al.* (2011) found that, irrespective of students' demography, school was by far the most popular source of information on sex. A survey of Australian young people (Giordano and Ross 2012) found that they wanted and expected a consistent approach to sex education in Australian schools, with every respondent seeing schools as important in sex education. According to Cohen *et al.* (2012), surveys from such different countries as Canada, Australia, United States and Tanzania consistently show that the vast majority of parents support broad-based sex education in schools.

On the basis of this literature, it would appear that many stakeholders believe that school-based sex education is valuable, and many believe that it should be broad-based and address issues of concern to adolescents. However, the evidence suggests that this is not happening consistently in the United Kingdom or a range of other national contexts. Effective delivery of unbiased, accurate, multi-faceted sex education within schooling contexts, replete with social sensitivities, is indeed a challenging goal to attain.

Sex education in schools: controversial content

Writing in 2000 in the context of the United Kingdom, Lynda Measor commented that sex education was one of the most controversial and politicised aspects of the school curriculum. Citing Trudell (1993), she went on to argue that it drew the attention of adults with distinct cultural, political and economic agendas who engaged in heated and bitter debates in which student voices are largely unheard. The controversy continues across developed western nations, such as the United States and Australia (Peppard 2008; Santelli *et al.* 2006); for example, between supporters of secular humanist and 'comprehensive' sex education, which informs students about a range of methods of contraception (including but not limited to sexual abstinence) and disease prevention, and the Conservative Right's platform of 'abstinence-only' sex education that has prevailed for many years in the United States.

In the United States, teachers therefore have found themselves in difficult positions with respect to sex education because of concerns that parents and school officials will not support their efforts. Consequently, according to Donovan (1998), teachers fear that discussion of controversial topics such as masturbation, sexual orientation, abortion and even contraception may jeopardise their careers. Irvine (2002) argues that conservative opponents of comprehensive sex education in the United States have scored an impressive political victory by paralysing countless community debates and constraining programmes nationwide, despite widespread support for sex education. By arguing sex education is radical, dangerous and immoral, they have fostered a climate in which scores of professionals have suffered reprisals for speaking in support of sex education.

These barriers to comprehensive sex education are not restricted to the US context. About a quarter of teachers in one study in the United Kingdom were unable to teach what they thought was appropriate sex education because of fears of negative reaction from parents (Fisher and McTaggart 2008). Similarly, just under 50 per cent of teachers surveyed for an Australian study said that they were careful about what they taught because of concerns about negative community reactions (Smith *et al.* 2011). The United Kingdom and Australia have also experienced pressure from abstinence-only lobby groups. For example, a recent private member's bill in the United Kingdom has promoted abstinence to girls, although oddly enough not to boys (The Public Whip 2011), and Christian Right groups and supporters strongly opposed a sex education programme in South Australia (Peppard 2008).

We agree with the views of commentators such as Santelli *et al.* (2006) that the 'abstinence only' position is morally problematic and threatens human rights. In seeking to persuade adolescents to abstain from sexual intercourse, it directly disregards:

> the right of all persons, free of coercion, discrimination and violence, to [among other things] decide to be sexually active or not [and to] pursue a satisfying, safe and pleasurable sexual life
>
> (World Health Organisation 2002)

In addition, it appears that abstinence-only policies are less effective. A comparative study across four nations including the United States (Weaver *et al.* 2005) suggests that sexual health indicators are better in the three countries with practical and comprehensive sex education than in the United States.

However, with the comprehensive sex education that we advocate, unanswered questions remain about what should be taught, by whom, and where should the line be drawn? What people think should constitute comprehensive sex education is clearly contested ground, and will vary enormously with the different cultural norms that exist within and between national boundaries. However, students' voices on these issues are now being heard, and it appears that most students want good, accurate and comprehensive education about sex and sexual relationships.

What students want from sex education

It seems from the available literature that at least in the Australia and United Kingdom most students favour a comprehensive approach to sex education that addresses their concerns. Writing in the context of the United Kingdom, Measor summarises the two main aspects of sex education where students find the information provided inadequate:

> The first relates to emotional content: 'They talk about your body, not about your emotions.' The second complaint concerns the lack of explicit information on a wide range of topics: 'I thought it wasn't really good enough for information about sex, oral sex, gays, bisexual people, etc.' Young people were critical that sex education did not deal directly

with sex and with the experience of sexuality. It failed to give explicit information about a number of topics and did not discuss alternative sexual orientations.

(Measor 2000: 122)

These themes reappear in a recent survey of over 1,200 15–29-year-old Australians (Giordano and Ross 2012). It found that, in addition to the usual topics of anatomy and reproduction, safe sex, STIs and contraception, young people wanted sex education also to include healthy relationships, sexuality, sex and pleasure, accessing youth health services and HIV/AIDS. The survey specifically recommended that sex education be inclusive of young people who are same-sex attracted and sex and gender-diverse.

Tinning (2004: 242) refers to an experienced head of a Health and Physical Education (HPE) department in an Australian secondary school who considered that the most frequent issues raised with her by students were 'relationships, peer pressure and sex'. Issues relating to pregnancy, the Pill, break-ups with boyfriends and negotiating peer group disputes were also dominant in her teaching experience. HPE teachers are more often the interface for students wanting to know about the social implications of sexual maturity than they are for fielding questions relating to the location of the fallopian tubes. However, Tinning (2004) claims that the reality is that issues of 'desire, pleasure and sensuality' are marginalised content in HPE.

Young people also have clear ideas of how they would like their sex education taught. Newby *et al.* (2012) found that amongst English secondary students, small group work and group discussions were the most popular methods of delivering sex education. These authors called for Sex and Relationship Education (SRE) 'incorporating a variety of teaching methods, recognising and expressing the potential for different points of view and levels of experience, ensuring that pupils are signposted to additional sources of information to meet differing levels of need' (249). North American adolescents identified the use of active, rather than didactic, teaching methods as important in Sexual Health Education (Eisenberg and Wagner 1997).

Australian young people were asked who they would like most to teach them sex education (Giordano and Ross 2012). Responses of participants in the study clearly indicated that mainstream teachers, whether from the disciplines of Personal Development, Health and Physical Education (PDHPE) or science, were not high on their list. Sixty-eight per cent of respondents nominated trained sexual health peer educators (preferably not too much older than the students themselves) as their preferred option for sex education in schools, and 68 per cent nominated sexual health educators from community organisations. Only 32 per cent nominated PDHPE teachers, while science teachers came in just ahead of faith-based organisations with only 19 per cent of respondents nominating them as a preferred option. In fact, 45 per cent of those surveyed did not want science teachers to deliver sex education content at all.

That these young people were against science teachers as preferred sex education providers accords with curriculum changes over the past two decades that, while broadening and extending the scope of sex education, have largely moved it out of the science area.

Sex education in school curricula in the United Kingdom and Australia: the move away from science

Sex education in UK curricula

In the early 1990s, the first author conducted some research into sex education at a comprehensive school in the Midlands of England, where he was then a science teacher. At that time, sex education was largely in the domain of science education and tended to be predominantly biologically focused (Taylor and Brierley 1992). The research indicated that this form of sex education was not addressing many of the issues then of concern to adolescents, or, as Jones (1989) put it, it failed to ease the considerable unhappiness caused to some young people as a result of the many taboos associated with sexual matters.

Since that time, sex education – now called Sex and Relationship Education (SRE) in the United Kingdom – has become part of a subject known as Personal, Social and Health Education (PSHE) (Qualifications and Curriculum Authority 2007a), with some limited biological aspects still mandated in secondary science curricula. However, PSHE developed as a non-statutory subject, so schools were not required to follow its programmes of study, giving schools the option to teach SRE according to their own systems of values.

In light of this, Buston *et al.* (2010) found that a lack of consensus on how much and what kind of sex education should be taught led to patchy provision across schools. Often the values, experiences and characteristics of individual classroom teachers were strong determinants of what sex education was actually delivered. The situation was further complicated because PSHE programmes have become increasingly crowded, have no set curricula and are not examined. These issues contribute to the problem of evaluating the outcomes of sex education effectively. In addition, although sex education has been compulsorily taught, parents have the right to withdraw their children from non-statutory (i.e. non-science) SRE lessons (PSHE Association 2010).

In the past few years, PSHE and SRE in UK schools have been reviewed several times, particularly focusing on its controversial non-statutory status. It was strongly argued by Fisher and McTaggart (2008) that PSHE should be a statutory subject containing core statutory SRE content, and that 'it is not good enough that some children and young people only receive the biological aspects of SRE that are covered in Science'. A subsequent report (Macdonald 2009) recommended that PSHE education should become part of the statutory National Curriculum, but that the parental right to withdraw students from SRE be maintained except for the limited sex education covered in the science curriculum. A government Bill (Department for Children Schools and Families 2010) announced that PSHE would be a new statutory National Curriculum subject in Key Stages 3 and 4, and that the parental right to withdraw children would operate only for children under 16. However, subsequently a departmental review outlined by the Chair of the Education Select Committee (Gibb 2011) claimed that the government had ruled out making PSHE statutory.

Hence the science curriculum (Qualifications and Curriculum Authority 2007b) contains the only inescapable sex education in UK schools. However, it covers limited aspects of sex education, such as human reproduction and sexual health including contraception and STIs. This is separate from the broader 'relationships' component of SRE which is taught as part of the PSHE non-statutory curriculum according to the policies of individual schools, and from which children under 16 can opt out.

Sex education in Australian curricula

Like the United Kingdom, prior to the 1980s, formal sex education in Australian schools was limited in scope and extent, and predominantly anatomical. Historically in the state of New South Wales, sex education was delivered by science teachers and private providers such as the Family Life Movement or the local community health nurse. In the late 1980s and 1990s, in part spurred on by the spread of HIV/AIDS in Australia, a national policy statement *Talking Sexual Health* (Australian Research Centre in Sex Health and Society 1999) advocated inclusive school-based education about sexuality and sexual behaviours. These issues were to be integrated into a broad health education curriculum, explicitly including the needs of homosexual students and students with disabilities. This policy was endorsed by all states and territories.

The strong political influence of university-based Health and Physical Education scholars influenced this shift of sex education into becoming a mandatory Health and Physical Education key learning area, modified in the early 1990s with the inclusion of Personal Development. Essentially, and in summary, in Australia there has been a shift of sex education from Science to Health and Physical Education.[1]

Hence in New South Wales, content relating to emotional and social aspects of puberty was introduced into the Physical Education syllabus, and subsequently into the mandatory Personal Development, Health and Physical Education (PDHPE) syllabus. At the primary school level, aspects of sex education, such as reproduction, puberty, menstruation, sexual identity and STDs including HIV/AIDS, are currently addressed entirely in the PDHPE syllabus, along with issues relating to gender images and personal identity; not in the science syllabus.

At the secondary level, issues related to sexuality, sexual health, pubertal changes and sexual identity are largely 'addressed' in Health and Physical Education, which does not cover the structure and function of human reproductive systems. This is the only area of sex education covered in the secondary science syllabus, via the relevant essential content point for Stage 5 students (typically 15–16 years old), which specifies that students learn to 'relate the organs involved in human reproductive systems to their function' (Board of Studies New South Wales 2003: 35).

This location of sex education outside the science syllabus is about to be further cemented as Australia implements its first National Curriculum. Sex education is one of the focus areas of the National Health and Physical Education (HPE) curriculum. Students in Australian schools, from Foundation (kindergarten) to Year 10, will be provided with developmentally sequenced learning experiences around sexuality and reproductive health within the Key Learning Area of HPE. Regular opportunities for teachers to revisit these focus areas from Foundation to Year 10 are connected to students' growing maturity and their increasing ability to understand more complex concepts (ACARA 2012: 22). As such, at least at the planning stage, the Australian HPE curriculum outlines sequenced and iterative focused learning in sex education.

In the National Curriculum, sexuality and reproductive health are defined broadly (within the focus area of HPE) as 'including understanding and managing physical, social and emotional changes that occur over time, exploring sexual and gender identity, managing intimate relationships, understanding reproduction and sexual health, and accessing community health services'. This sequencing is in alignment with the proposals of Health Literacy (Nutbeam 2008), which is a hallmark of the Health Education approach found in the Australian curriculum's area of HPE. Health literacy is defined as the ability to recognise, understand and effectively manage emotions, and use this knowledge when thinking, feeling and acting (ACARA 2012: 30).

The learning continuum included in 'The Shape of the Australian Curriculum: Health and Physical Education – August 2012' outlines the broad learning sequence (13–16) from Foundation – 'basic understanding of how their body is growing and changing' – to what 'makes them similar and different' – to 'manage the physical'... leading to 'physical, emotional and social changes associated with the start of puberty' progressing towards 'recognising sexual feelings and evaluate behavioural expectations for different social situations' (ACARA 2012: 18). In contrast, the Australian National Science Curriculum contains very few relevant learning outcome statements. At the primary level, these are limited to understanding general growth and development of living things. At the secondary level, the outcomes refer to structure and function considerations of organs and systems in living things.

So, similar to the situation of UK students, for Australian secondary students, the 'biology of sex' remains for the foreseeable future located within the Science Key Learning Area and separate from the relationships and social and emotional factors associated with sex education, which are the focus of HPE. However, unlike the United Kingdom, Australian students cannot opt out of the sex education or any other component of the formal HPE curriculum.

Sex education in schools: controversial quality

Given that it has been suggested that the majority of young people gain their sexual health education knowledge from school, and that this is provided mainly by teachers (Walker 2001), it is unfortunate that teacher-delivered sex education is often not well taught and does not provide young people with the understanding deemed necessary to negotiate sexual relationships (Wright *et al.* 2002).

In the United Kingdom, reviews of SRE by the Office for Standards in Education (Ofsted) indicated that, in secondary schools, teaching about sexual health was inadequate, including teaching about sexually transmitted infections and the law in relation to sex. As well, non-specialist teachers were often reluctant to contribute to all topics of SRE (Ofsted 2002). A survey of over 20,000 UK youth aged under 18 (UK Youth Parliament 2007) also painted a grim picture: 40 per cent of respondents thought their sex education was 'poor' or 'very poor', while 57 per cent of girls aged 16–17 had not been taught how to use a condom and 43 per cent of students not taught about personal relationships. Fisher and McTaggart (2008: section 17) summarised the findings of a survey of young people's views on UK SRE, criticising the subject as 'not relevant to young people's real lives; it was not given sufficient curriculum time; it was delivered by untrained teachers; and it was not inclusive of LGBT young people and young people with disabilities'. The parallel teacher survey showed that teachers were also concerned about the quality of SRE, particularly aspects related to relationships.

Clearly some of these problems are likely to relate to the lack of dedicated curriculum time for teaching SRE in many schools, as identified by Ofsted (2010). However, much deeper and more complex constraints are also at play. Buston *et al.* (2010) found that those teachers heavily involved in guidance and pastoral care in UK schools usually delivered PSE and hence sex education. They argue that the main motivation in taking on a pastoral care role was unlikely to be a desire to deliver sex education, and many may not be comfortable with that aspect of their role. Furthermore, many teachers worried about imposing their views and values on students.

When teachers feel under-prepared for teaching or uncomfortable with the content of sex education, this can manifest itself in a bookish, didactic pedagogical response, as reported above by DiCenso *et al.* (2001). The comfort level of individual teachers to speak about physical, emotional and social aspects of sex and sexuality, enter into supportive partnerships with parents, and adapt to the cultural and social values and attitudes of individual schooling contexts will influence the level of effectiveness of sex education for primary and secondary school students (Milton 2003). Wilson-Gahan (in Callcott *et al.* 2012) highlights the all-too-common position of health education as 'little more than a wet weather activity' (160), with health education for students comprising 'handouts of labelling parts of the reproductive system' (163). In Australia, a survey of sex education teachers indicated that their teaching focused on the negative aspects of sexuality, with the pleasure aspects of sexual activity being taught by less than 50 per cent of respondents (Smith *et al.* 2011).

In both the United Kingdom and Australia, and further afield, a consistent and strong message is that some of these problems are associated with a lack of appropriate training. This problem arises from both inadequate instruction during undergraduate preparation and from a lack of staff development and training opportunities once teachers enter the classroom.

Preservice teacher education and professional development in sex education

In the United Kingdom, Buston *et al.* (2010) found that the lack of mandatory training and limited opportunities to attend courses on the delivery of sex education meant that those who felt uncomfortable with the area had little opportunity to gain confidence. Westwood and Mullan (2007) pointed to the lack of full-time PSHE teachers in most UK schools, and Forrest *et al.* (2002) argued that leaving PSHE to form tutors places them in an awkward position in terms of possible embarrassment and confidential disclosure. A review of SRE in UK schools (Fisher and McTaggart 2008: section 18) identified lack of teacher training as 'the most significant barrier' to effective SRE teaching and outlined a list of recommendations specifically aimed at improving the skills and confidence of those who deliver SRE.

Further afield, similar quality concerns surround teacher preparation for sex education, for example as raised by Donovan (1998) in the US context. According to Cohen *et al.* (2012), many Canadian teachers were being asked to teach Sexual Health Education without adequate preparation, and many did not feel willing to provide this, while others were willing to cover some topics, but not all. Moreover, when required to provide Sexual Health Education, these teachers often did not use interactive teaching methods nor did they encourage questions (DiCenso *et al.* 2001; Byers *et al.* 2003a; 2003b; Meaney *et al.* 2009). It appeared that, even if these teachers were using a well-resourced and supported Sexual Health Education curriculum, they would be unlikely to provide experiences that had positive effects on students' sexual decision-making and sexual health outcomes. A survey of Australian secondary school teachers of sexual education (Smith *et al.* 2011) found that 16 per cent of respondents had no training at all in sex education and most relied on very limited one-off or short-term professional development opportunities. The teachers felt they had inadequate access to professional development in this area, and the more sensitive the topic (for example, sexual abuse or same sex attraction), the more support they needed.

Conclusion: implications for science teaching

Our starting point for this chapter was a holistic and inclusive view of sex educa-tion that does more than present negative approaches, disease and dysfunction, but rather promotes positive dimensions of sexuality and relationships. Given this aim, is the science classroom the place for this kind of sex education?

From a practical and pragmatic perspective, our answer is 'no' – that the moves of sex education away from science that we have seen in the United Kingdom and Australia are inevitable and appropriate. One of the respondents to a survey of Australian youth claimed that 'anatomy should stay predominantly in science, dis-cussions about respectful relationships don't belong in a science classroom' (Giordano and Ross 2012: 24). In some ways we agree with this view. Teachers of this kind of sex education need to be very well prepared for the complexities of teaching and learning related to this sensitive area, and need to be willing and able to navigate the diverse family, community and religious agendas that surround it. In addition, they need to be aware of potential mismatches between their values and the prevailing values of the teaching context. For example, should a teacher favour-ing a comprehensive approach follow this line if employed at a conservative school? What about the converse situation? Science teachers, particularly beginning science teachers, are well out of it and have more than enough challenges in dealing with some of the other controversies and debates that are outlined in this book.

However, science teachers do need to be appropriately prepared to deal with the challenges of the limited sex education for which they are also responsible. In contexts where science curricula still include contraception and STIs, this is particularly important. In a student-centred classroom environment that encour-ages open questioning and discussion, it is very likely that this area of science teaching will intersect with broader issues of behavioural choices and relation-ships. But should science teachers gag or should they foster any emerging classroom discussions that move into the syllabus territory of PDHPE/PSHE? Certainly these teaching moments offer considerable scope for powerful teaching and learning that promotes sexual health in its widest sense. But these opportunities bring with them the risk of triggering objections from individual students, par-ents, school governors or the broader community, depending on the context. This risk also applies to teaching methods: how about using a zucchini as a prop to demonstrate how to put on a condom? Good practical hands-on teaching to some, but possibly highly offensive to others.

A beginning science teacher's ability to negotiate this tricky territory will be enhanced by a good understanding of the policy frameworks and values of their schools, and some prior reflection on teaching methods and how to respond to questions that might be received. Pre-service science teachers, and the students they go on to teach, would benefit from formal teacher preparation for this aspect of teaching, together with appropriate professional development.

Although we welcome the broadening of sex education and the associated moves away from the purely biological and anatomical, we also consider that stripping the broader elements of sex education out of science curricula weakens the teaching and learning of the relevant science. Constructivism, currently the predominant theoretical framework of science teaching and learning (Tobin 1993), emphasises the importance of engaging students through issues relevant to them for constructing scientific understandings. Broad issues of sexuality and sexual relationships are of enormous relevance to adolescents, so uncoupling the biological and other scientific concepts from these issues of relevance would result in a less powerful teaching and learning experience.

Moreover, an emerging literature highlights the educational advantages of connecting scientific information with broader social issues (Zeidler *et al.* 2005). This literature argues that, if scientific knowledge is located within students' social, political, and cultural lives, then students will develop 'functional scientific literacy', enabling them to make informed decisions incorporating moral and ethical principles.

Separating the science from the broader issues related to human sexuality and relationships is a backward step from this perspective. This is evident from a clear inconsistency between the Australian National Curriculum Science content and its framing paper (National Curriculum Board 2009), which states as one of its goals that 'students should be able to ... make informed decisions about their own health and wellbeing'. Students will not be helped to make informed decisions on their sexual health on the basis of the limited relevant content of the science curriculum alone.

What much of the preceding discussion leads to is the need for teaching sex education in partnership. A recurring theme in the literature surrounding school-based sex education is that sex education should not be taught in isolation: it needs to be linked to other parts of the curriculum and make use of appropriate external providers (Fisher and McTaggart 2008). We argue that science educators have an important specific role to play in any sex education partnership. A scientific and evidence-based perspective is required to contribute appropriate, accurate terminology together with clear and unambiguous teaching about relevant biological structure/ function concepts and causal connections between sexual behaviours and consequences. Science educators can enhance the teaching of their science as well as the broader sexual education of their students by creatively exploring cross-curricular opportunities to link the science of sex with the broader issues of sexual health and relationships that are so important to adolescents and to society.

Key questions

1   Having read this chapter, reflect upon your approach to or views about sex education. Do these reflect a holistic and inclusive view of sex education?
2   Now reflect upon the type of sex education that you received. Was that holistic and inclusive?
3   Has this chapter in any way changed your views about how sex education should be presented to children?

Further reading

Donatelle, R. J. (2013) *Health the basics* (10th edn). San Francisco, CA: Pearson.

Mayle, P. (2000) *Where did I come from? A guide for children and parents.* New York: Kensington Publishing.

Meldrum, K., and Peters, J. (2011) *Learning to teach health and physical education: The student, the teacher and the curriculum.* Frenchs Forest, NSW: Pearson Education.

Note

1 Personal communication, Professor Jan Wright, University of Wollongong, 30 August 2012.

References

ACARA (Australian Curriculum, Assessment and Reporting Authority) (2012) *The Shape of the Australian Curriculum: Health and physical education – August 2012.* Available online at: www.acara.edu.au/verve/_resources/Shape_of_the_Australian_Curriculum_Health_and_Physical_Education.pdf (accessed 20 August 2012).

Australian Research Centre in Sex Health and Society (1999) *Talking Sexual Health: National framework for education about STIs, HIV/AIDS and blood-borne viruses at secondary schools.* Report developed for the Australian National Council for AIDS, Hepatitis C and Related Diseases (ANCAHRD). Victoria, Commonwealth of Australia: La Trobe University.

Bearinger, L. H., Sieving, R. E., Ferguson, J., and Sharma, V. (2007) Global perspectives on sexual and reproductive health of adolescents: Patterns, prevention and potential. *The Lancet*, 369(9568): 1220–31.

Board of Studies New South Wales (2003) *Science Years 7–10 Syllabus, Minor amendments incorporated July 2009.* Sydney: Board of Studies New South Wales.

Buston, K., Wight, D., and Scott, S. (2010) Difficulty and diversity: The context and practice of sex education. *British Journal of Sociology of Education*, 22(3): 352–68.

Byers, E. S., Sears, H. A., Voyer, S. D., Thurlow, J. L., Cohen, J. N., and Weaver, A. D. (2003a) An adolescent perspective on sexual health education at school and at home: I. High school students. *Canadian Journal of Human Sexuality*, 12(1): 1–17.

Byers, E. S., Sears, H. A., Voyer, S. D., Thurlow, J. L., Cohen, J. N., and Weaver, A. D. (2003b) An adolescent perspective on sexual health education at school and at home: II. Middle school students. *Canadian Journal of Human Sexuality*, 12(1): 19–33.

Callcott, D., Miller, J., and Wilson-Gahan, S. (2012) *Health and Physical Education: Preparing educators for the future.* Melbourne: Cambridge University Press.

Cohen, J. N., Byers, S. E., and Sears, H. A. (2012) Factors affecting Canadian teachers' willingness to teach sexual health. *Sex Education*, 12(3): 299–316.

Department for Children Schools and Families (2010) The Children, Schools and Families Bill. Reference number DCSF-01098-2009. UK: Department for Children, Schools and Families. Available online at: http://www.publications.parliament.uk/pa/cm200910/cmbills/008/10008.i-iii.html (accessed 5 November 2013).

DiCenso, A., Borthwick, V. W., Busca, C. A., Creatura, C., Holmes, J. A., Kalahian, W. F., and Partington, B. M. (2001) Completing the picture: Adolescents talk about what's missing in sexual health services. *Canadian Journal of Public Health*, 92(1): 35–8.

Donovan, P. (1998) School-based sexuality education: The issues and challenges. *Family Planning Perspectives*, 30(4): 188–93.

Eisenberg, M. E., and Wagner, A. (1997) Viewpoints of Minnesota students on school-based sexuality education. *Journal of School Health*, 67(8): 322–6.

Fisher, J., and McTaggart, J. (2008) *Review of Sex and Relationship Education (SRE) in Schools*. A report by the External Steering Group. London: Department for Children, Schools and Families.

Forrest, S., Strange, V., and Oakley, A. (2002) A comparison of students' evaluations of a peer-delivered sex education programme and teacher-led provision. *Sex Education*, 2(3): 195–214.

Gibb, N. (2011) *Review of personal, social, health and economic education (PSHE)*. UK: Department of Education. Available online at: https://www.gov.uk/government/news/review-of-personal-social-health-and-economic-education (accessed 5 November 2013).

Giordano, M., and Ross, A. (2012) *Let's talk about sex: Young people's views on sex and sexual health information in Australia*. Surrey Hills, NSW: Australian Youth Affairs Coalition (AYAC) and Youth Empowerment Against HIV/AIDS (YEAH).

Goldman, J. (2008) Responding to parental objections to school sexuality education: A selection of 12 objections. *Sex Education*, 8(4): 415–38.

Goldman, J. (2012) A critical analysis of UNESCO's International Technical Guidance on school-based education for puberty and sexuality. *Sex Education*, 12(2): 199–218.

Irvine, J. M. (2002) *Talking about sex: The battle over sex education in the United States*. Berkeley, CA: University of California Press.

Jones, R. (1989) *Personal and social education: Philosophical perspectives*. London: Kogan Page.

Jones, T. M., and Hillier, L. (2012) Sexuality education school policy for Australian GLBTIQ students. *Sex Education*, 12(4): 437–54.

Macdonald, A. (2009) *Independent Review of the proposal to make Personal, Social, Health and Economic (PSHE) education statutory*. UK: Department for Children, Schools and Families. Available online at: http://www.educationengland.org.uk/documents/pdfs/2009-macdonald-pshe.pdf (accessed 5 November 2013).

Meaney, G. J., Rye, B. J., Wood, E., and Solovieva, E. (2009) Satisfaction with school-based sexual health in a sample of university students recently graduated from Ontario High Schools. *Canadian Journal of Human Sexuality*, 18(3): 107–25.

Measor, L. (2000) *Young people's views on sex education*. London: RoutledgeFalmer.

Milton, J. (2003) Primary school sex education programs: Views and experiences of teachers in four primary schools in Sydney, Australia. *Sex Education*, 3(3): 241–55.

National Curriculum Board (2009) *Shape of the Australian curriculum: Science*. Canberra, ACT: Commonwealth of Australia.

Newby, K., Wallace, L. M., Dunn, O., and Brown, K. E. (2012) A survey of English teenagers' sexual experience and preferences for school-based sex education. *Sex Education*, 12(2): 231–51.

Nutbeam, D. (2008) The evolving concept of health literacy. *Social Science and Medicine*, 67, 2072–8.

Ofsted (Office for Standards in Education) (2002) *Sex and relationships: A report from the Office of Her Majesty's Chief Inspector of Schools* (HMI 433). London: Ofsted.

Ofsted (2010) *Personal, social, health and economic education in schools*. London: Ofsted.

Peppard, J. (2008) Culture wars in South Australia: The sex education debates. *Australian Journal of Social Issues*, 43(3): 499–516.

PSHE Association (2010) *Sex and relationships education guidance to schools (Consultation) DCSF 2010 (existing DFES 2000): Guidance document in PSHE education guidance: A summary of government guidance related to PSHE education*. PSHE Association. Available online at: http://www.pshe-association.org.uk/resources_search_details.aspx?ResourceId=320 (accessed 5 November 2013).

Qualifications and Curriculum Authority (2007a) *PSHE: Personal wellbeing: Programme of study (non-statutory) for Key Stage 3*. Available online at: http://www.education.gov.uk/schools/teachingandlearning/curriculum/secondary/b00198880/pshee/ks4/personal/programme (accessed 5 November 2013).

Qualifications and Curriculum Authority (2007b) *Science: Programme of study for Key Stage 3 and attainment targets*. Available online at: http://www.fondation-lamap.org/sites/default/files/upload/media/minisites/EIST/PDF/science_2007_programme_of_study_for_key_stage_3.pdf (accessed 5 November 2013).

Santelli, J., Ott, M., Lyon, M., Rogers, J., Summers, D., and Schleifer, R. (2006) Abstinence and abstinence-only education: A review of US policies and programs. *Journal of Adolescent Health*, 38(1): 72–8.

Smith, A., Schlichthorst, M., Mitchell, A., Walsh, J., Lyons, A., Blackman, P., and Pitts, M. (2011) *Sexuality Education in Australian Secondary Schools 2010*. Results of the 1st National Survey of Australian Secondary Teachers of Sexuality Education, Monograph Series No. 80, Melbourne: Australian Research Centre in Sex, Health and Society, La Trobe University. Available online at: http://www.latrobe.edu.au/arcshs/news/the-first-national-survey-of-secondary-teachers-of-sexuality-education (25 June 2013).

Taylor, N., and Brierley, D. (1992) The impact of the law on the development of a sex education program at a Leicestershire comprehensive school. *Journal for Pastoral Care and Personal and Social Education*, 10(1): 23–9.

The Public Whip (2011) Sex Education (Required Content) Bill. Available online at: http://www.publicwhip.org.uk/division.php?date=2011-05-04&number=266&display=allvotes&sort=vote (accessed 4 May 2011).

Tinning, R. (2004) Rethinking the preparation of HPE teachers: Ruminations on knowledge, identity, and ways of thinking. *Asia-Pacific Journal of Teacher Education*, 32(3): 241–53.

Tobin, K. (ed.) (1993) *The practice of constructivism in science education*. Hillsdale, NJ: Lawrence Erlbaum Associates.

Trudell, B. N. (1993) *Doing sex education*. London: Routledge.

UK Youth Parliament (2007) *SRE: Are you getting it?* London: UK Youth Parliament.

Walker, J. (2001) A qualitative study of parents' experience of providing sex education for their children: The implications for health education. *Health Education Journal*, 60: 132–46.

Weaver, H., Smith, G., and Kippax, S. (2005) School-based sex education policies and indicators of sexual health among young people: A comparison of the

Netherlands, France, Australia and the United States. *Sex Education: Sexuality, Society and Learning,* 5(2): 171–88.

Westwood, J., and Mullan, B. (2007) Knowledge and attitudes of secondary school teachers regarding sexual health education in England. *Sex Education,* 7(2): 143–59.

World Health Organization (2002) *Defining sexual health.* Report of a technical consultation on sexual health, 28–31 January 2002. Geneva: World Health Organization.

Wright, D., Raab, G. M., and Henderson, M., Abraham, C., Buston, K., Hart, G., and Scott, S. (2002) Limits to teacher delivered sex education interim behavioural outcomes for randomised trial. *British Medical Journal,* 324: 1430–6.

Zeidler, D., Sadler, T., Simmons, M., and Howes, E. (2005) Beyond STS: A research-based framework for socioscientific issues education. *Science Education,* 89: 357–77.

Chapter 15

The mystery of the body and the laboratory

Steve Alsop

Introduction

My title seems to lend itself to an all-too-familiar literary genre: the tyrannical scientist conjuring life back from beyond the dead; or a post-mortem scene within a crime scene investigation; or, even, Professor Plum with the dagger in the conservatory. In this light it is, perhaps, an altogether curious way of starting to think about science education. Yet, I believe that this line of enquiry opens up a series of germane and pressing considerations. I actually feel it rather important to reflect on what it means to have a body (or not), physical or otherwise, in education. This becomes even more intriguing, I suspect, within a subject such as science that seems to treasure its inspection and objectivity. I discuss all of this later. For now, I start with two provocations, both in the form of questions. How ought we think about the body in science education? What body might we care for within our practices?

An exploration should, perhaps, start with the obvious. Without our bodies we are unable to see, to move, to touch, to taste and to hear. We cannot breathe and our hearts cannot raise a beat (or, for that matter, even beat at all). Indeed, we are unable to outwardly or inwardly express any thoughts or feelings. Without our body, with all its complex and situated nuances, there are no possibilities of experience and therefore no possibility of education at all. Possibly even more daunting, of course, there is no prospect of life at all. And yet when I write this seemingly obvious point, this manner of thinking seems somehow to be elusive and bizarrely abstract, especially in science and education. As science educators, we all know that we have bodies. So to spend any time dwelling on them seems a rather hapless and jovial pursuit; one certainly not warranting protracted academic scrutiny, sensibilities or temperament. Perhaps other than a source of amusement, what possible difference can talk of the body make?

In what follows, my gathering argument is that we should actively engage in far more 'Body Talk' (Latour 2004). After all, our bodies are both the very subject and object of *all* our educational experiences. My intent is to first uncover the mystery of the missing body. Although this is widely overlooked,

I consider it to be a significant debate in science and science education. In my early discussions, I will seek to shed light on the corporeal vanishing act in academic culture. The following sections attempt what might be described, somewhat dramatically, as an awakening, or an exhumation of sorts, in which I seek to revive the body as a central analytical concern and concept in science education. Let me begin.

Science incarnate

Christopher Lawrence and Steven Shapin edited a delightful collection of essays on the topic of relationships between the body and the body of scientific knowledge (Lawrence and Shapin 1998). A central point, made by all the assembled authors, is the historical trick of the vanishing corpus – the body of science seems to have no body at all, and yet all those involved were paradoxically alive and fully embodied (if you see what I mean). In an opening chapter, Lawrence and Shapin openly wonder about our propensity of recounting richly-embodied experiences in disembodied and transcendental scholarly ways. A tradition, in which they note:

> … intellectuals have enshrined their products, and the objects of their inter-
> pretations in the transcendental and disembodied domain that knowledge
> itself was understood to inhabit. Such versions of where knowledge was
> located did not need to be justified; in much academic discourse they have
> been taken for granted. In dominant sensibilities, therefore, the response to
> the mystery inscribed in the relationship between embodied knowledge-
> making and disembodied knowledge itself was just not to talk about it.
> Indeed, not talking about it has been a way of ensuring its continued mys-
> terious status.
>
> (Lawrence and Shapin 1998: 2)

So this seems to be our first clue. Dominant academic sensitivities inscribe disembodied knowledge to represent richly-embodied experiences. As an edu-cator, I am subsequently intrigued by how scientists and others are socialised into accepting such sensitivities, and the ways in which our common science education practices are implicated (or not) in this process. I wonder whom this tradition might serve well, as well as those it might not.

The separation of the mind from the body has a long-standing and fre-quently-rehearsed history in western philosophical thought. The elevation of the mind as the essence, spirit and personification of humanity is likely to have complex origins. Many, however, trace its origins back to ancient Greece. Within *The Republic*, Plato outright dismisses Homeric poets and their appeals to ephemeral, unreliable and untrustworthy feelings and emotions. In this light, the body is perceived as a residue of erratic and unpredictable emotional life; in contrast, the mind, *cognito*, is the crux, the core, of pure reason and rationality.

The mind is ethereal and divine, as near to God as is humanly possible. The body, in complete contrast, is of mere instrumental worth – a fleshy mechanical and material entity. While stories of Rene Descartes' mind–body duality are often a little more mythical than historical records suggest (see Dear 1998), within associated 'Cartesian' narratives, the body and the mind seem to be easily separated and naturally typecast in opposition (Alsop 2005).

This image is found widely. For instance, within contemporary societies there is an ever-present imaginary of a genius as somebody processing superior brainpower (more especially in science and mathematics). To search for the mystery of Einstein's sources of brilliance in his body is most definitely bizarre. Indeed, some might even say warped or even perverted. There is a quirky BBC documentary that traces Professor Kenji Sugimoto's pursuit of Einstein's Brain. In the opening sequence, Professor Sugimoto, a professor of mathematics and science history at the Kinki University of Osaka Japan, is filmed explaining to a US passport officer that the purpose of his visit to the United States is: 'Looking for Einstein's brain'. Although this is somewhat eccentric in itself, one hesitates to imagine what would happen if Professor Sugimoto were to announce a comparable fascination for Einstein's body. The backdrop for the documentary is that, shortly after Einstein's death, a highly respectable and curious pathologist at the University of Pennsylvania removed Einstein's brain. Small pieces were then distributed to a variety of experts, with somewhat unclear goals (as far as I can work out). His body, in contrast, was cremated. It remains an open debate to this day whether this was Einstein's actual desire.

The associated mystery of Einstein's brain, the laboratory in Pennsylvania and Professor Sugimoto's quest is, alas, another story. The point I am making here is, of course, illustrative. Einstein's brain continues to be a source of our fascination and intrigue as it is seen as the source, or essence, of his brilliance. His body – in complete contrast – is of far less importance. The familiar image of a Roswell-like Alien Being basically makes the same point – we readily associate advanced beings as having enlarged heads, bulging eyes and emaciated bodies (cf. movie characters ET and more recently 'Paul'). It almost seems that intellectual advancement necessitates some type of eye, mind and body trade-off.

A question that is rumbling beneath these discussions concerns relationships between knowledge and the process of knowledge-making itself. This question might be cast within the slightly different context of education and reframed in more familiar educational terms. What are relationships, we might ask, between our conceptual understandings of science (our knowledge and thoughts) and our embodied educational experiences? Or phrased slightly differently, what are relationships between learning science and our corporeally embedded experiences of science education. As the essays in Lawrence and Shapin's (1998) collection make abundantly clear, that while there is a dominant academic convention of disembodying ideas, in the ways in which these are inscribed, recorded and presented, there is an intriguing line of exploration which takes up the task of reassembling and reviving the bodily presence in knowledge-making processes. This retelling and reframing offers an intriguing historical enquiry into science knowledge production and the ways in which experiences garner intellectual status as they jettison the mundane, corporeal and contextualised aspects of their developmental origins. The manner in which scientific knowledge makes claim to authority is – in part – built on the ways in which it obviates the contexts and thus associated contingencies of its production.

The body of science education

Let us now turn our attentions to science education. Within school science practices, students are socialised into the conventional norms of academic disciplines. Our places of learning in this respect serve not only to learn science concepts and skills, but also to legitimise and delegitimise ways of being in science that include ways of thinking and feeling as well as ways of sensing (seeing, touching, smelling, hearing and tasting – although taste is understandably rarely directly encouraged). The rituals or *rites of passage* into science have familiar and common trajectories that promulgate more and less appropriate ways to think-act-feel, as Sara Delamont and colleagues make delightfully clear (Delamont *et al.* 1988). Science and science education in this regard is a socialising process in which some ways of being are constituted, while others are supressed and occasionally even open to reprimand. As part of this process, school science socialises students into performing academic and school-based traditions of disembodiment. Here are a few 'cherry picked' examples.

It is not uncommon, for instance, to see an abundance of photographs of whole people in elementary school science teaching materials. In high school, however, these start to morph into more abbreviated and sketchy line-drawn representations. The classic suspended 'eye' in optics is perhaps a prime example. The body in biology is also increasingly represented in objective, anatomical, impersonalised drawings. Even textbooks of biology that focus on the body are intent on parcelling it out into free-floating biological systems that seemingly function autonomously and independently from each other with presumably degrees of self-determination. Such systems include the digestive system, the nervous system, the respiratory system, the immune system and a host of other systems as well. As authors in Science and Technology Studies bring to our attention, through such processes, the human body becomes a naturalised object of medical enquiry that is almost machine-like in form and can be laid out on a laboratory table and understood through objective study. In this manner it seems that even the study of the human body in science and biology education is in danger of losing some of its holistic form.

It is widely recognised that the literary genre of science and also science education tends to remove 'superfluous' contextual references. The way we encourage children to record their science investigations is, perhaps, a case in point. The traditional laboratory style reports focus on accurately recording procedures, data collected and reporting any conclusions drawn. It would be unusual, to stay the least, to make extended note of: how comfortable the seating was; or how tricky it was to manipulate the apparatus ('boy, my back was killing me during that fiddly titration'); or the lighting, décor or ambiance of the setting; or indeed any other incarnate quotient experiences. The sensibility is usually to remove any notion of the person performing the particular experiment.

We also have a tendency to account for our teaching and learning practices in a similar way. If you were to pick up a midterm school report card at random, you are likely to find comments about learning and understanding such as: 'John has clearly understood forces and motion'; or, 'Jane has excelled at astronomy and has grasped the concepts of a black hole and stellar cycles'. You will be much less likely to read accounts of John and Jane's physiology, dexterity, postures or habits. Indeed, their body is more likely to be associated – if mentioned at all – with inappropriate rather than appropriate acts such as hyperactivity.

The body often appears when teachers discuss practical work and laboratory skills. Indeed, there is a curious expression 'hands-on and mind-off' that is sometimes used to convey a lack of balance between the mind and body, usually referring to hazards of too much time spent on manual activities and not sufficient time mentally engaged with theories or concepts. Our lesson plans, in this respect, frequently seek an implicit degree of balance between different types of learning activities: a long, potentially inactive and immobile lecture is perhaps best tempered with an accompanying more dynamic and energetic group exercise, or a role-play or practical work. An appropriate balance is likely to increase student engagement. In the context of these discussions, what is perhaps interesting to note is the way that the mind-and-body become once more cast in opposition (to return to an earlier point). It is also intriguing to ponder whether it is ever possible to have one 'on' while the other is 'off', and visa versa, much like the figures entering and exiting a weather clock.

Academic research in science education also fails to give any sustained attention to the embodied nature of teaching and learning. We have a long-standing interest in a Piagetian epistemic entity, and decades of constructivist-inspired research on Children's Ideas in Science have firmly established the fallacy of the empty knowledge vessels, or *tabula rasa*. Conceptual change is now quite readily represented as an active process in which learners construe and construct meaning based on existing understandings or residing cognitive structures. Our preoccupations in this regard have been with 'knowledge-entities' and 'knowledge vessels', but not really with holistic experiential forms. Jean Piaget, in an early book I recall, offers an eloquently engaging story of a walk during which his son is observing the moon. This is an instance that is used to illustrate egocentric reasoning. The fact that they are outside at night walking seems somewhat irrelevant and superfluous – a detail of peripheral significance. Of late, our attentions in science education have taken a more sociocultural turn. Social and cultural features now seem to be self-evidently important in understanding teaching and learning, which might be attributed to the efficacy and proliferation of studies with orientations of this type in our field. Yet the idea that learners are more than social and cultural manifestations seems irrelevant. Where are, and how ought we think about their bodies, I almost hesitate ask? The general point is that our disposition in science education is to account for education in terms of dominant psychological and sociological narratives and not in terms of the body per se.

These arguments seem even more salient within discussions of pedagogy. Try to imagine teaching without a body? It is comical, to say the least. Yet, most of our discussions of pedagogies tend to focus on language and curricula content, such as analysing Socratic questioning or the type and style of argumentation, to give two popular examples. Classroom conversations and exchanges are deconstructed in terms of their linguistic content, as well as socio-linguistic and semantic structures. The actual location and production of these utterances, however, seems altogether rather unimportant. If bodily gestures are mentioned at all, it is usually to accompany and illustrate some primary aspect of language. We seem much more comfortable talking about education in terms of disembodied language exchanges rather than dwelling on 'seemingly redundant' details, such as where language comes from in the first place.

As I have previously mentioned, the body is commonly associated with feelings and emotions. I have had a long-standing interest in the study of emotions in science education. It is a topic with a 30-year-plus history. But while we remain fascinated by studies of 'attitudes-toward-science', often recorded by Likert-scale type psychological tests completed outside of lived educational settings, we pay much less attention to emotions and their effects during the lived *experiences* of science education. Emotions are demonstrably a central part of our daily practices. It is impossible to conceive of a classroom (or any other living event) that is emotion-free. In research, however, it is almost as though we value post-event rationalisations of emotions to the near exclusion of these situated reactions. In contrast, I continue to believe that daily-lived situated emotions are fundamental to the efficacy of science education in both the short and longer term (see Alsop 2005).

A host of post-modern and feminist theorists bring to our attention the idea that 'in so many ways we *are* our bodies. Thus, we might think of the body's outward appearance as a social skin of sorts', to use Lesley Sharp's words (Sharp 2007: 7). The 'body politic', of course, has been the focus of so much powerful social theorising.

I would like to thank a colleague Brian Mathews for his words of caution concerning the use of the term 'body', as it is a dualistic statement of sorts. There is an inherent danger, as I have noted elsewhere (Alsop 2011), of starting with assumptions that separate the mind and body and then tautologically argue for unification. The notion of 'embodiment' also has associated weakness, as Brian notes, because it is so individualistic, and thus fails to fully acknowledge the socially, materially and temporally *situated* natures of our bodies. We always reside somewhere, and our being in the world, in this regard, cannot be framed as independent. The title of my chapter should – perhaps – more appropriately refer to the *self and the laboratory*, in recognition of our mimetic 'fleshy' insertions into the social world and associated interactions, rationalities and reciprocities.

My preference here for the term body draws its origins from Lawrence and Shapin's (1998) essays and other selected phenomenological work within Science and Technology Studies and Environmental Education. The term also offers multiple opportunities to connect with dominant narratives and imaginaries within science and science education (such as the body in the laboratory). The body, rather than the self, is something that science has laid claim to, and this makes it intriguing. So, I use the concept of the body whilst acknowledging some of the weakness that this entails. Incidentally, there is a reciprocal tension of losing discussions of aspects of the body (such as physiology, morphology, sensations and affect) in some social and cultural narratives. In *being* a body we experience joy, happiness, sadness, anguish and pain (as well as a gamut of other sensations and emotions). In this regard, the 'body politic', as described in so much social theorising, is 'sensually' thin, it is left underdeveloped as a 'theoretical body' that 'never aches', or 'suffers under the elements', or 'experiences fear', 'exhilaration' or 'muscle strain', as Rebecca Solnit (2001: 28) so delightfully brings to our attention.

Tracking down the body

Up to this point, I have been deconstructing the notion of the body within cultural traditions of science and science education. I want to now turn my attention to a more reconstructive agenda: the prospect of bringing the body back to life in school science and in the school science laboratory. Given the weight of our academic culture is often furthered by disembodiment, as you might imagine, this is far from straightforward. Especially given that the ways that we commonly account for our practices follows popular academic cultural traditions.

My previous discussions have suggested practices that serve to advance disembodiment, and these each offer points of reflection and explorations on how these might be changed and performed differently. We could, for instance, encourage more explicit references to emotions, feelings and aesthetics within our classroom activities. We could more purposefully plan for such events, and conceive of ways of describing and reporting our lessons with these in mind. We could change our scientific inscriptions. Laboratory write-ups do not need to be disembodied, and neither does our 'eye' in optics or 'systems' in biology. These could all be reassembled through discussions and different types of pedagogical representations, enactment and performances. We could get our bodily acts together.

We could also start to think about our practices in terms of embodiment. While there are cultural traditions that we need to be sensitive to, we are not compelled to represent science or science education as disembodied. One response might be to try to place greater attention on lived experiences with and within science and our practices of science education. We probably readily associate experiential education with outdoor settings or wilderness retreats. Bearing this in mind, it likely seems rather odd to contemplate experience as a mode of knowledge and knowing in more traditional settings, especially in setting such as a science laboratory. But perhaps the body of science education that we ought to care more for is an 'experiential body'; after all, our bodies carry us around and they are clearly always present whether we are in the field, forest, garden or in the lab. The point is that we *experience* in all settings, and yet we seem somehow intent in accounting for these settings in different experiential ways.

This is a line of argument that I am certainly sympathetic with. I argue that we need to listen much more closely and more intently to the aesthetic, emotional and ethical aspects of our daily and our students' daily encounters with science. These can too easily become obscured by the ways in which we have learned to talk about academic knowledge and knowing, and measure and record academic progress. In contrast, I am intrigued to explore what it might mean to *embody* educational practices in school science contexts.

Dissecting a body part

In a recent study, Arvola Orlander and Per-Olof Wickram bring attention to rambunctious teenagers and their bodily encounters with secondary science (Orlander and Wickram 2011). They describe two quite different contexts: conducting a calf-eye dissection and learning about sex. As you might imagine, such experiences are going to be rather lively, as the following extract from one of their classroom observations (ibid.: 14) reveals:

Helen: I won't be able to cope with this. I will …
Jonna: Well, if it's going to smell like things, then I'm leaving at once.
Helen: If it smells, then I'll leave. You know if it smells, shit.
Jonna: No, oh shit, now I feel sick.

Helen and Jonna's visceral reactions to eye dissections serves to highlight that the body (much like the mind) is far from empty, it is not a *tabula rasa*. In being living bodies, Helen and Jonna discern/construct/experience (there are a number of different words here) the putrid smells and jellied eye flesh simultaneously with visceral disgust, revulsion, anguish and nausea. These learners, in this way, are actively involved in negotiations of bodily experiences as learning science, as Orlander and Wickram note. Although such reactions are highly unlikely to be part of the assessed curriculum, they seem so central to any meaning-making because they constitute the 'very context in which meaning-making takes place' (Orlander and Wickram

2011: 24). The body in this sense is a context for all our educational experiences. Given this, how can we not take seriously Helen and Jonna's constituted embodiments within their scientific encounters with bodies of others?

What stands out in Orlander and Wickram's (2011) study, however, is a deep-rooted pedagogical contradiction. Helen and Jonna's predictably lively reactions contrast so starkly with the expectations and reactions of their teacher who finds these reactions unacceptable. As we can fully appreciate, some topics in science are going to be lively. But in this instance, while such reactions are to a certain extent predictable, they are also undesirable and unsuitable in some ways. I recognise that these reactions might be affecting the studying of others, but, along with Orlander and Wickram, I suggest that this is only part of the issue. In accordance with the cultural academic traditions of science, the teacher feels somehow compelled to repress or smother overly-lively emotional reactions. They simply don't seem right in some way, or they are over the top, or they might unduly upset others. Whatever it is, this is certainly not suitable behaviour for a science laboratory. In this respect, I suggest that Helen and Joanna's classroom teacher is confronted with a dilemma of how to negotiate gaps of what culturally ought to be occurring, and what is occurring – navigating, what I have called elsewhere, the 'politics of embodiment' (Alsop 2011).

Orlander and Wickram's (2011) study records an awaking of the body within a context in which it is deemed in some way as unnecessary. This example helps us to start to conceive of political 'body-work' in science and education. These teenagers in all their liveliness and ebullience help us to recover a view from somewhere, a location that is always present, interpretative and structuring; a body (cf. Haraway, 1991). On some occasions, of course, this becomes more noticeable, on other occasions less so.

I wonder what it means to learn to listen more closely and intently to the body-work of science education? Perhaps one way of starting this process is to learn to tolerate and understand our emotional encounters and offer them a more central role in how we think about and perform science education. If we learn to accommodate such heightened emotions, we might be better positioned to recognise, nurture and support other more subtle bodily encounters. Who knows, we might also start to plan for a more embodied pedagogy that explicitly and purposefully invites – even calls for – particular ethical, aesthetic and emotional encounters and meaning-making.

What if science education went wild?

There is a delightful essay by Anthony Weston (2004) in which he poses the question: What if teaching went wild? The challenge that he sets himself is to find ways of working towards a 'radically different practice and philosophy of (environmental) education within schools as we know them' (ibid.: 31). He continues by noting that a wild pedagogy is 'much more personally demanding and unnerving than the usual sorts of pedagogical innovations' (ibid.). The goal of

Anthony Weston's pedagogy is Earth awareness and Earth responsiveness, and he draws on Paul Shepard's (1982) notion of the lost development of self as an account of contemporary environmental destructiveness. Weston's response is to set an agenda for change that involves reconnecting with the Earth through 're-acknowledging ourselves as animals' and actively sensing our presence as 'parts of larger living systems'. The setting is a school classroom, which he makes clear is often unsuitable for this task.

Weston's first challenge is to become aware that we ourselves are actually present. One way of beginning this inquiry, he suggests, is to ask a class or small group to pack themselves into half the teaching space – such that pupils becoming aware of each others' presence. During this process, students start to re-awaken themselves as embodied beings within a particular teaching and learning context. Weston then asks individuals to look closely at their hands and to write a report on questions that include: 'What can you tell about this person just from looking at their hand – Sherlock Holmes style, as it were? Where has this hand been? What is the person's occupation? How dry is the weather? ...' (Weston 2004: 7). A few minutes, he proposes, is enough to notice the distinctive, unique details of your hands and how they bear hallmarks of nature and nurture, your genetics and lived histories. They also have many similarities in shape and mechanics with other non-human animal hands (as Weston makes clear). Following this, Weston asks learners to look closely at a neighbour's hands in much the same way as theirs, recording and discussing details, asking and answering questions. Even this straightforward activity, he suggests, completely changes the ambience and energy in the classroom. It loosens things up and makes them more open. He reflects on his experiences of teaching this activity. At the end of the activity, he writes, 'now people let go of their hands, pull their seats a little bit apart. Even so there is a remembrance of embodiment that lingers, something people can carry away and think about' (ibid.: 38). Students over the years, he recalls, have come back to him repeatedly and said how much they 'liked the hand thing' (ibid.: 38).

Weston continues the theme of touch with an invitation to hold in your hand for some time a familiar local artefact, such as a stone or twig, and a rather distant object, such a meteorite (that is, older than the Earth itself). He then encourages students to touch a flower and contemplate the contrasting texture, softness and smell of the flower. The haptic experience of holding your hand, or somebody else's hand, or holding a rock, or flower offers a context for discussion of embodiment and relationships, as well as the ways that we are all deeply connected when we touch. Weston then invites students to eat their flowers (chosen in advance to be of an edible kind). The idea of eating a flower seems quite unnerving for some students and raises discussion of what it is to eat something, to digest something and incorporate it into 'ourselves'. In another example, he invites students to look around the classroom, to locate, inspect life and speculate where they might be living and what they might be doing. He now mentions to the class that he has brought in some insects, a Daddy Long-legs and the class start to react, to wriggle and squirm a little. As Weston makes clear, this is not a show or tell, or a standard science experiment, but rather a phenomenological and philosophical awaking to attend to how it 'changes our sense of space, when we discover such Others already present' (ibid.: 44).

So, what does it mean for teaching to go wild? A practical classroom-based answer, Anthony Weston suggests is:

> to have a sense […] it is to recognise that even the shape or our own aware- ness (e.g. our own animality) often eludes us. Wild is that unsettling sense of otherness, unexpected and unpredictable and following its own flow, but still a flow that is, in some not-quite-graspable way, ours too.
>
> (Ibid.: 45)

I recognise that the activities that Weston suggests might seem distant from traditional practices in science education, especially in secondary school. How- ever, the idea of planning for embodiment and recognising that our awareness, in this regard, often eludes us, offers intriguing referent points for practices. Would it be possible to approach science education with similar goals in mind? If so, what would these practices and experiences entail?

Conclusion

My argument has been about knowledge, education and the body. It is an unde- niable feature of academic cultures that the body seems of much less importance than the mind, particularly in high-status subjects such as science. Indeed, it has been repeatedly noted that learned cultures elevate their status by disregarding the everyday, mundane and bodily aspects of their pasts. My intent has been to put questions of the body on the epistemic and pedagogical agenda, and to raise questions of how we ought to think about the body in science and science educa- tion. As well as a supplementary question of what body might we care for (or not) within our practices.

With considerable assistance from Christopher Lawrence and Steven Shapin's historical studies of science in the making (Lawrence and Shapin 1998), Arvola Orlander and Per-Olof's Wickram's study of Helen and Jonna dissections (Orlander and Wickram 2011) and Anthony Weston's invitation to go wild, we have re-encountered and re-awakened the body as a source of awareness and meaning-making. We are fully aware that the living body is an *object* of medical and biological study (it is measured by various medical tests and scans; it com- prises of complex systems, DNA and cellular structures). The body object lies still on the laboratory bench for dissection. But, perhaps even more importantly, the body is also a *subject* as well. As human beings, as learners and teachers, we are fully embodied. Our bodies *are* us. Without them, as I mentioned in my opening paragraph, there is no prospect of experience, or education, or life at all. In *being* bodies, we experience the world and come to understand, sense and feel the world in diverse ways, including scientific ways.

Bruno Latour (2004), with his particular writing style, outlines a course in Paris where students can be educated to smell. The pedagogy comprises of interactions with a perfume odour kit that consists of different fragrances of various types and intensities. During the course, participants are slowly able to learn to be more attentive to different chemicals found in perfumes. They literally become a nose, and they learned to discern and interpret smells in new ways, as Latour explains. One of my colleagues, Natasha Myers (2006) studies X-ray crystallographers and how they learn to make sense of and manipulate complex molecular structures. Her ethnographic study records the scientists drawing on their bodies as interpretative sources and for communication. The ability to manipulate complex equipment and interpret and manipulate dense graphical virtual computer images of protein molecules draws on our corporeal experiences with real-world artefacts. Protein crystallographers' molecular embodiments offer a wide range of new questions for social studies of science, as Myers (2006) concludes by noting. It also, I believe, has profound implications for science education as well.

The general point that I am making is that through our diverse experiences within science education we come to make meanings in particular ways. The education of learners' bodily senses is an undeniable part of meaning-making and thus science education. In this regard, within our existing practices, we already care for and constitute a particular *scientific* body, although we rarely acknowledge and account for this. I leave open for another occasion the normative question of whether this is the body we ought to (or indeed need to) care for? As well as how we might start to care for this body with greater intentionality.

My discussions have sought a modest agenda: to provoke more 'Body-Talk' (Latour 2004) in science and science education. Now, more boldly, I assert that I believe there to exist a whole line of enquiry that might be built around such talk, which to date has only been tangentially explored. So my closing provocation is an invitation to make a 'mind-shift', well, actually, a body-shift – or perhaps an altogether more unified shift of *self* – and explore your practices and knowledge of science as increasingly diverse and efficacious embodiments. Why not follow Anthony Weston's examples and let your science education go wild? I suspect that many of your students will find this exceedingly appealing, if not entertaining and a tad amusing.

While academic traditions of portraying scientific knowledge as disconnected might help to elevate such knowledge, they can also create discontinuities between our real worldly situated embodied experiences and practices and the immaterial and rarefied knowledges that we study. Identifying and helping learners safely navigate this gap is a key agenda for science education. It is also a gap that many learners can find intimidating and off-putting (Alsop 2005).

In closing, my response to 'The mystery of body and the laboratory' is to ask ourselves (as science educators and researchers) how we could reawaken our (and others') actual presence in science education. In so doing we can help learners (and ourselves) experience the world in ever-new ways, by becoming increasingly influenced by others. Framed in this way, the goals of science education might now even be conceived as the development of an increasingly *scientifically embodied body*. This is a challenge that I leave you with.

Key questions

1   How ought we think about the body in science education?
2   What body might we care for within our practices? Why? And how?
3   What would it mean to be more attuned to the emotional and moral experiences of teachers' and learners' bodily encounters in science education? (Orlander and Wickram 2011)
4   What does, and ought it, mean being human in science education?

Further reading

Sharing an interest in embodiment and phenomenology, this chapter draws from a number of different research fields. Those intrigued might enjoy reading the following.

Within the Science Education literature

Orlander, A., and Wickram, P.-O. (2011) Bodily experiences in secondary school biology. *Cultural Studies of Science*, 6: 569–94.
Alsop, S. (2011) The body bites back. *Cultural Studies of Science Education*, 6: 611–23.
Wickram, P.-O. (2006) *Aesthetic experience in science education*. London: Lawrence Erlbaum Associates.
Alsop, S. (ed.) (2005) *Beyond Cartesian dualism: Encountering affect in the teaching and learning of science*. Dordrecht: Springer.
Perrier, F., and Nsengiyumva, J. (2003) Active science as a contribution to the trauma recovery process: Preliminary indications with orphans from the 1994 genocide in Rwanda. *International Journal of Science Education*, 25: 1111–28.

In different ways, these publications raise interesting questions about embodiment, emotions and morals within science education.

Within the Science and Technology Studies literature

Lawrence, C., and Shapin, S. (1998) The body of knowledge. In C. Lawrence and Shapin, S. (eds) *Science incarnate: Historical embodiments of natural knowledge*. Chicago, IL: University of Chicago Press (pp. 1–20).
Myers, N. (2006) Animating mechanism: Animations and the propagation of affect in the lively arts of protein modelling. *Social Science Studies*, 19(2): 6–30.
Puig de la Bellacasa, M. (2011) Matters of care in technoscience: Assembling neglected things. *Social Studies of Science*, 4(1): 85–106.

From different perspectives, these publications explore relationships between scientific bodies of knowledge and (scientific) bodies.

Within the general Education and Environmental Studies literature

The chapter draws from an phenomenological perspective, with touches of eco-phenomenology. In this regard, the reader's interest might be piqued by:

Abrams, D. (1996) *The spell of the sensuous.* New York: Vintage Books.
Abrams, D. (2011) *Becoming animal: An earthly cosmology.* New York: Vintage Books.
O'Loughlin, M. (1997) Corporeal subjectivities: Merleau-Ponty, education and the postmodern subject. *Education Philosophy and Theory*, 29(1): 20–31.
Weston, A. (2004) What if teaching went wild? *Canadian Journal of Environmental Education*, 9: 31–46.

These engaging publications all foreground embodiment as a pressing theme for education research and practice.

References

Alsop, S. (ed.) (2005) *Beyond Cartesian dualism: Encountering affect in the teaching and learning of science.* Dordrecht: Kluwer Press.
Alsop, S. (2011) The body bites back! *Cultural Studies in Science Education*, 6: 611–23.
Dear, P. (1998) A mechanical microcosm: Bodily passions, good manners, and Cartesian mechanisms. In C. Lawrence and S. Shapin (eds) *Science incarnate: Historical embodiments of natural knowledge.* Chicago: University of Chicago Press (pp. 51–82).
Delamont, S., Benson, J., and Atkinson, P. (1988) In the beginning was the Bunsen: The foundations of secondary school science. *Qualitative Studies in Education*, 1(4): 315–28.
Haraway, D. (1991) Situated knowledges: The science question in feminism and the privilege of partial perspective. In D. Haraway (ed.) *Simians, cyborgs, and women: The reinvention of nature.* New York: Routledge (pp. 183–201).
Latour, B. (2004) How to talk about the body? The normative dimension of science studies. *Body and Society*, 10(2–3): 111–34.
Lawrence, C., and Shapin, S. (eds) (1998) *Science incarnate: Historical embodiments of natural knowledge.* Chicago, IL: University of Chicago Press.
Myers, N. (2006) Animating mechanism: Animations and the propagation of affect in the lively arts of protein modelling. *Social Science Studies*, 19(2): 6–30.
Orlander, A. A., and Wickram, P.-O. (2011) Bodily experiences in secondary school biology. *Cultural Studies in Science Education.* doi:10.1007/s11422-010-9292-4.
Sharp, L. (2007) *Bodies, commodities and biotechnologies.* Columbia: Columbia University Press.
Shepard, P. (1982) *Nature and madness.* San Francisco, CA: Sierra Club.
Solnit, R. (2001) *Wanderlust: A history of walking.* London: Penguin Books.
Weston, A. (2004) What if teaching went wild? *Canadian Journal of Environmental Education*, 9: 31–46.

Chapter 16

World science, local science

Peeranut Kanhadilok

Introduction: what is 'World Science' and what is 'Local Indigenous Science'?

For the sake of this chapter World Science, or Western Modern Science, is the science that we all know and love: it is that system of 'knowledge-making' that seeks to answer questions about the nature of the physical world we inhabit, from the macro to the micro. The Arts and Humanities Research Council's definition of science is as follows:

> Science is the pursuit and application of knowledge and understanding of the natural and social world following a systematic methodology based on evidence.
>
> (AHRC 2012)

It is a valued system of knowledge that relies on certain laws established through the application of scientific methods to phenomena in the world around us. Very generally, the processes of scientific methods begin with an observation followed by a prediction or hypothesis that is then tested. Depending on the results, the hypothesis gains validity, can be tested further, gains reliability and can, eventually, becomes a scientific theory or 'truth' about the world. The history of Western Modern Science (WMS) demonstrates that it developed in Europe, in particular over the last 150 years. Its 'truth' relates to certain values and ideas, and is not necessarily objective. That said, although scientists may admit that there are many ways of understanding the natural world, they believe that science is the best way, because of its testable knowledge.

On the other hand, indigenous knowledge (or Local Indigenous Science, LIS) in this chapter is a term that includes the beliefs and understandings of some western but mostly non-western people – knowledge that is acquired through long-term association with particular places. The term indigenous knowledge is meant to convey 'a complete knowledge system with its own concepts of epistemology, philosophy, and scientific and logical validity' (Dais quoted in Battiste and Henderson 2000: 41). It is knowledge based on the social, physical and spiritual understandings that have informed people's survival, and has contributed to their sense of being in the world. Indigenous science goes by many different names, such as Traditional Ecological Knowledge (TEK), Indigenous People's Knowledge (IPK), and even folklore or 'folk knowledge'.

If we understand 'indigenous' to relate to people who have a long-standing and complex relationship with a local area and 'science' to mean a regular approach to acquiring knowledge of the natural world, then indigenous science is the process by which indigenous people build their empirical knowledge of their natural environment. Indigenous knowledge is not just about people in cultures other than the West: there are many examples of 'home-grown' knowledge in the United Kingdom and elsewhere. This might range from the simple admonishments a mother might make ('Don't go out with your hair wet, you'll catch a cold!', 'Don't sit on a cold step, you'll get piles!') to 'old wives' tales', farming and folklore about weather changes when crows gather or cows lie down in fields, and onwards to villagers in Somerset and their ideas about the nature of radiation and radioactivity in an area of naturally high background count (Alsop 2005).

Indigenous knowledge systems are the sum total of diverse streams of knowledge gathered over time, accepted by the community and synthesised into collective knowledge as necessary for survival. Traditional teachings are provided for learners to integrate elements within this process themselves; it has been called a process of 'cultural breathing' (Aikenhead 2006). Individual and collective kinship patterns, experiences, relationships, interpretations of community practices, spirituality and history are all important factors that contribute to cultural epistemology. Knowledge is drawn in, internalised and expressed by individuals and by collectives. As is the case with WMS, LIS is the practical application of local theories of knowledge about how things work in the world. The question at the heart of this chapter is whether these two forms of science/knowledge should ever be integrated and merged, whether people should incorporate western scientific knowledge increasingly within their everyday local practices. In this chapter we discuss the interactions between WMS and LIS, although, in our case, we relate this in particular to the local indigenous science of just one country: Thailand.

What is the relationship between Local Indigenous Science and Western Modern Science?

The similarities and differences between local indigenous and western science have received a great deal of attention in the literature (e.g. Aikenhead 2001; Cajete 2000; Hammond and Brandt 2004; Michell 2005; Snively and Corsiglia 2001; Stephens 2000). Writers like these generally distinguish between scientific knowledge and indigenous knowledge by claiming that science is universal whereas indigenous knowledge relates only to particular people and their understanding of the world. There are occasions when science takes up some aspect of indigenous knowledge but only when this meets the criteria of western science. One approach is that science and indigenous knowledge represent two different views of the world around us: science focuses on the component parts whereas indigenous knowledge presents information about the world in a very holistic way. Generally, however, local indigenous knowledge does not fit the criteria for western science and therefore is classed as a different – often inferior – kind of knowledge. There is no doubting that WMS has enormous power, is seen to be 'imperialistic' and can dominate (disparage, dismiss, extinguish) local knowledge.

Table 16.1 Western Modern Science and local Thai wisdom compared

Western Modern Science	Local Thai wisdom
Universal or global knowledge This is knowledge generated in modern scientific institutions or industrial firms. It has 'universal truth' regardless of location.	*Local knowledge* Local wisdom is rooted in particular communities. It is a set of experiences generated by people living in those communities and is often specific and context-bound.
Explicit This knowledge has been explicated rigorously during the processes and procedures of creation, through observation, experimentation and validation.	*Often implicit and tacit* Local wisdom is embedded in people who generate and use it. So, it is often difficult to capture and classify this kind of non-formal knowledge.
Transmitted in written form through academia and education WMS is created and carefully documented. It is commonly transferred by teaching in formal education systems.	*Transmitted orally and culturally* Local wisdom is rarely recorded in written form. It is mostly commonly transferred through imitation or demonstration with cultural contexts.
Theoretical knowledge Knowledge is derived from hypothetico-deductive methods and scientific procedures. Studies are commonly made in laboratories, with scientific or mathematical models.	*Practical and experiential rather than theoretical knowledge* Local wisdom is generated from experience, trial and error. It is tested through time in the 'social laboratory of survival' within local communities.
Objectivist values The knowledge creation process is separated from attitudes, beliefs and cultural dimensions.	*Founded within religious worldviews, spirituality and social values* Spirituality is an important and inseparable dimension of local wisdom. Nature is often revered as the 'mother' or provider of all things.
Compartmental approach The process of knowledge breaks down matters of study into smallest components in order to reach into deeper and hidden facts of what is being studied.	*Holistic approach* Humankind is considered part of nature. There is a nature tendency toward equilibrium. It is the central theme of local wisdom.

Source: Adapted from Tinnaluck (2005).

For science teachers and educators, there are four key approaches to the relationship and interaction between the two: (1) to work to fully integrate them; (2) to fight to keep them separate; (3) find some kind of compromise in saying that *some* aspects of WMS and LIS might actually come together while the rest remain detached; or, of course, (4) to rule out LIS altogether and just teach western science. This is the debate at the heart of this chapter: should LIS be taught at all, and, if so, should LIS and WMS be integrated and taught as such (in part or in whole), or kept distinct and disconnected?

The first question to answer is: Why teach LIS at all? There are two main reasons. First, by introducing students to the concept of indigenous knowledge in their science education, they will have an increased awareness of their own culture and identity, and those of other cultures. Second, to understand that modern day environmental problems have social and cultural dimensions which benefit from perspectives other than just western science. While scientific knowledge is needed to solve these problems, science alone is often not sufficient, and indigenous science may make a useful contribution. It is possible, then, to see how both systems can inform, and how one system might complement the other.

The cross-cultural perspective is popular, and is seen as an issue of political fairness and equal opportunity. Minorities are seriously under-represented in the sciences, so the goal here is to improve primary and secondary education, so that all students have an equal chance to make progress in the sciences if they so choose. However, Ogawa (1995) warns that this approach is not a simple matter. Indigenous science is more than just a single knowledge system comparable to western science. Instead, it is 'a body of stratified and amalgamated knowledge and cosmology' (ibid.: 1). For this reason, 'A simple [curricular] transformation from one system to another is not often feasible' (McKinley 2005). It is more than just finding eminent scientists from other countries to humanise science lessons.

The other viewpoint is that the indigenous science of, say, Australian aborigines, Canadian First Nation Inuits or New Zealand Maoris, is a valid way of knowing and learning about the natural world. Different cultures have powerful ways to address critical environmental, health and energy problems that threaten the earth's communities. Therefore, the argument goes, the school science curriculum should integrate traditional values, teaching principles and concepts of nature intermixed with those of WMS. The worry here, according to Aikenhead (2006) is that 'integration can fail depending on the context' (ibid.: 123). It is difficult to see how the two can easily integrate and, clearly, there is no simple formula for full cross-cultural science education.

In this debate, we place ourselves with the camp that recommends keeping both knowledge systems strong – and yet separate. Garroutte (1999) published a paper entitled *American Indian Science Education: The Second Step* that provides a synthesis of the contrasting assumptions in western scientific and

American traditional thought systems, and the importance of keeping these two models of inquiry fully detached when designing the science curriculum. He argued that it is possible to teach science, keeping the two models of inquiry distinct, without having to coerce one set of knowledge claims to fit the other. In Australia and New Zealand, such models of co-existence are often called 'two-way learning' (Ritchie and Butler 1990) and in the United States have been called 'bi-cultural instruction' (Cajete 2000; Kawagley 2000). Not a simple option, as Battiste (2000: 202) suggests: 'Creating a balance between two worldviews is the great challenge facing modern educators'. However, this is the place we want to go.

Thai wisdom masters and western science

Thailand, the kingdom of legends, is one the most 'fantasial' countries in the world. Its full and rich cultures and traditions are frequently told but often under-explained. It is a country with very unique, strong cultural traditions, and her peoples have largely Buddhist religious beliefs. These two factors are interlocked and affect all aspects of Thai life – including education. In general, Thai people lead peaceful, simple lives, for example, growing rice to eat, weaving cloth to wear, carving wood for household implements. They develop specialist skills and knowledge, often acquired directly from other craftspeople, weavers, textile-makers, potters, artists and fabric designers, who make up the communal environment in which they live (Office of SMEs Promotion 2010). Older citizens have an important role in Thai society, and contribute to the family organisation and the community because of the respect accorded to them, their wisdom, wealth of experience and transfer of knowledge to younger generations. There is a tradition of transferring knowledge and experiences from and to people this way, without payment or fees. Many gain knowledge and experiences through this means, through informal learning at the feet of previous generations. Traditional Thai education has evolved to complement and sustain this unique and diverse culture. Thai local wisdom is one of the hidden stories that have been the pride of the kingdom for centuries. Sungsri (2011), for example, has stated that local wisdom is available in every province of Thailand.

Thailand's 1999 National Education Act, section 23, promotes traditional Thai wisdom within the educational system, such that it is intended to contribute to learning processes and knowledge, morality, religion, art and culture. The Office of the National Education Commission (2011) has defined Thai local wisdom as a body of knowledge, abilities and skills of Thai people that accumulates over many generations, through many years of experience, learning and development. There is a tradition of transferring knowledge and experiences to people, from generation to generation, often through local 'wisdom masters' (*Kru Pum Panya*) who are the philosophers, or wise men, of the community. Tinnaluck describes local wisdom as:

... accumulated experience rooted with a true understanding of nature, of which self or humanity is also a part. Wisdom is insightful knowing of everything based on natural law. Wisdom can occur only through collected experiences and a mind trained in consciousness.

(Tinnaluck 2005: 143–4)

Here we have been using the term 'Thai wisdom' not only because it is the closest translation of '*phoom panya*' ('*phoom*' means local, and can also mean insightful and smart, '*panya*' meaning knowledge) but also as a reflection of how Thai people perceive this age-old knowledge, i.e. that it is at the highest form of knowledge bound, to geographical or a community's context, suited best for each locality. Thai communities take pride in the term 'local wisdom' and there has been a resurgence of status of local wisdom in modern Thai life (Tinnaluck 2005). Ackoff (1989) places wisdom at the pinnacle of a hierarchy of knowledge, or 'knowledge pyramid', with five layers: (1) data, which comes from the raw observation and measurements; (2) information, which is created by analysing relationships and connections between the data, and is capable of answering simple questions such as 'Who?', 'What?', 'Where?' and 'When?'; (3) knowledge, which is the appropriate collection of information, and answers the question 'How?' and is a local practice, or relationship, between what works; (4) understanding, that answers the question 'Why?'; and (5) wisdom, which is created through using knowledge, through the communication of knowledge to users, and through reflection.

For Ackoff, wisdom is *permanent*, and he notes that, while data, information, knowledge and understanding relate to the past and deal with what has been or what is known, only wisdom deals with the future because it incorporates vision and design. With wisdom, individuals can create the future rather than grasp the

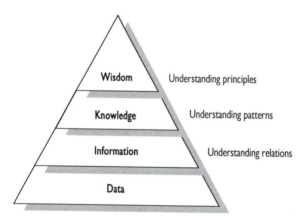

Figure 16.1 Ackoff's knowledge pyramid.

Source: Based on Akoff (1989).

present and past. Accumulated wisdom like this is not only used as a template for daily life, relating to families, neighbours, villages and their surrounding, but also to wealth-creation (Kongprasertamorn, 2007). Between 1998 and 2002, Thailand's Social Investment Fund project was launched within all 72,000 villages across the country, with a national scheme for providing one million Baht for each village. This project required all villages to focus on the production of one specific product each, and was called OTOP: One Tumbon (village) One Product. The aim was to promote the economy of each community in order to revive the overall economy of the nation (Tinnaluck 2005). The project was based upon King Blumibol Adulyadej's philosophy of sufficiency economy, and was intended to help address development challenges. This philosophy emphasises the way Thai people rarely separate the environment from human activities, and seek a balance in their interactions with nature: the balance between what humans take from the environment, and what they return, an equation which reflects upon their human well-being (Taylor *et al.* 2009). This philosophy promotes the use of local wisdom in the community as a guideline to developing people's work or careers. At the same time, the strategy encourages people to adopt modern science and technology to develop and improve their products.

In our own review of the nature of Thai local wisdom (Kanhadilok 2011), we see that its characteristics incorporate:

- *Stories* of, and within, local communities (Pungkanon 1992).
- *Outcomes of the collected experiences* through learning about the relationship with the environment, the relationship within a community and between different communities. This knowledge includes the conceptual of supernatural (Na Talang *et al.* 1995).
- *Directly and indirectly solving problems* in daily life through collected experiences (Pongpit 1994).
- *A body of knowledge, ability, and skills* that Thai people derive from their collection of experiences, their roots in Buddhism, Thai culture and belief (Office of the National Education Commission 1999).
- *Knowledge about values, morals and ethics,* respect for ancestors, spiritual practices and nature (Seri and Nantasuwan 2002).

More than 300 Thai wisdom masters have so far been honoured, playing an important role in developing human resources, and reshaping society. Each year, on 16 January, Thailand celebrates National Teachers' Day, with various ceremonies and activities, presided over by the Prime Minister, to show gratitude to teachers.

Western Modern Science and Local Indigenous Science in Thailand

The National Science Museum in Bangkok features a Traditional Thai Technology gallery, one of six major galleries, in this case designed specifically to inspire learning about the relationship between scientific knowledge and Thai local wisdom (National Science Museum 2007). One part of the gallery houses particular exhibits of metallurgy, including aspects of the moulding of metal that illustrate, for instance, the 'lost-wax process'. This process is one used to make sculptures of the Buddha, make complex and intricate ornaments of silver and gold. Some aspects of the process have transferred over a thousand years, for example, evidence of ancient lost-wax processes can be found in Ban Chiang village, Udantani Province, the north-east of Thailand.

Intricate works can be achieved by this method, primarily depending on the carver's skills. The basic technique begins with a wax model of the finished piece. Wax is easy to work with and inexpensive, and allows the sculptor to create a wax pattern of his or her very own. The wax model is covered by rubber latex that hardens. The wax is then heated and allowed to flow from the rubber mould and that then allows molten metal to be poured in to form a copy of the delicate wax shape. In Thailand, such intricate silverware is called 'Tom', where Tom means to 'put into'. In modern industrial uses, the process is called investment casting.

While Tom silver and goldware are beautiful and famous world-wide, the artisans in Thailand have clearly gained technological skills from western countries to develop their technology for the production of better goods, greater quantity and quality. With Tom gold, for example, craftsman pour the molten gold into pattern lines that can be quite deep from the surface of the product. Here an important technique is to use mercury paint on the surface of product before painting on the gold and producing the colour black (made from copper, lead

Figure 16.2 Thai goldware.

and silver melted together) to be black line of the pattern. Why mercury? Because mercury helps the gold be absorbed into the line design. Thai people have gained knowledge of such technology from production methods in western countries.

Within the Thai economy, Wua Lai Village has now become a community famous for its traditional silverware, helping to conserve the nation's artistic and cultural heritage. The art of silverware began in the region in the distant past, when forefathers made silver bowls and trays for use in religious ceremonies. The rhythmic sound of hammer on metal has been heard within Ban Wua Lai province for more than 200 years. One silversmith, Phrathan Bunlom, for example, was taught how to make silverware by his parents, knowledge passed from generation to generation within communities. In their book *One Tambon, One Product*, The Office of Small and Medium Enterprise Promotion (OSMEP) say:

> the silversmiths of Ban Wua Lai remain true to their traditions while adapting their skills and designs to the changing times, proudly bringing the ancient art of metal-working into the twenty-first century.
>
> (OSMEP 2012)

One of the OSMEP stories concerns Somsri Pewon, a third generation silversmith who recalls watching her parents and grandparents shaping and decorating raised patterns on silver bowls. Each silversmith retains the family's trade secrets and, in her school breaks, Somsri helped work alongside other family members to etch designs on the silverware. She is an advocate of progress, and argues that current production must respond to shifts in trend and fashion, with modern tools and techniques helping to develop the work. But to what extent do the new methods and old cultures intermix? While the new generation of young workers must be good at using new technology, says Somsri, they must also fully embrace tradition and heritage for the designs they use.

Another section of this gallery holds a collection of old traditional toys; a hands-on gallery where children and adults can play with the exhibits. The toys are made from local or waste materials that adults and children have

Figure 16.3 Thai goldware bracelet.

made and used in the past, and so reflect local wisdom, ways of life and the cultures of the community. The museum has organised elements of this gallery, called Traditional Thai Toy Activities (TTTA), that allow participants to experience pleasure by learning-through-doing in making and playing with the toys. This uses toys based on elementary scientific principles that closely simulate real-life scenarios, give scope for innovation and challenge, and make learning both LIS and WMS playful and exciting (Carruthers 2011). Nearnchalearm (2005), for example, points out that playing with traditional Thai toys promotes strong and positive attitudes toward both science and local culture, and makes coherent links between the two forms of knowledge and understanding.

In this first example, a 'coconut mouse' toy is made from coconut shell with string, a rubber band and wooden wheel. It falls in the category of force and motion toys. The mechanism is shown in Figure 16.4.

Thai culture emphasises relationships within the family. Thai families are usually large in number. It is not uncommon for at least three generations to live together in one house, for example, grandparents, parents and siblings. Grandparents will care for and educate their grandchild when mother and father work on the farm. The older generation will teach Thai culture to their grandchild through stories, the making and playing with toys, the relating of family history. Evening meals are communal, enjoying dinner together as a large family. To this extent, the coconut mouse toy is a traditional toy based upon traditional Thai wisdom. It is a toy encountered over generations; people have known how to make it and how to find the materials to construct it.

In the Thai traditional calendar, the first year of the animal cycle is the year of the mouse. Thai tradition holds that when all the animals of the world raced across the jungle to the finish line, the mouse came first, the cow second, then the following animals, the pig last. From this reason the mouse is the first of

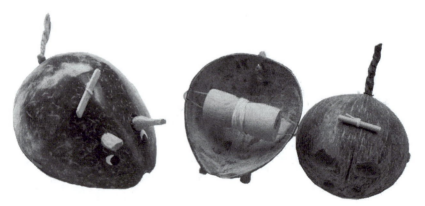

Figure 16.4 Coconut mouse toys.

the animal year cycle. Thai and Asian people believe that if the calendar turns to the golden mouse year, they will be wealthy. The animal year cycle represents the relationship between Thai people and animals in their life: mouse, cow, snake, goat, chicken etc. (Boonlert 2012). The mouse features in numerous Thai proverbs, for example: 'when the cat leaves, the mice enjoy', referring to a fairly universal proverb of when the employer is gone, the employees relax. 'The mouse falls into the rice pail' refers to someone poor who marries a very rich person; 'walking like a mouse in the trap' describes the action of the person moving towards trouble. Thai seafarers believe that when they see many mice run about the ship, it predicts a storm is coming (Yoksan 2012). On the other hand, farmers believe that mice are pests, not least because mice always eat and destroy products in their farms. It is not that farmers believe the mice are dirty; their belief is that these rodents will always eat the good products on the farms. That said, many traps have been created to rid themselves of the nuisance of mice. Then, from mousetrap to food, when the farmers get the mice in the trap, they are killed and prepared as food. The special mouse dish in Thailand is 'mouse curry'.

In terms of WMS, the toy moves from the spring of an elastic band, which covers the wheel and is fixed between the sides of the coconut mouse. It provides opportunities to consider the physics of linear movement, potential energy and kinetic energy. There can be discussion of the biology of the mouse, its role in the local ecology, that the mouse is the second or third consumer in the food chain. Mice are highly sensitive and there are tales that they are the first to scurry away at the very first signs of an earthquake. The biology of the coconut can also be provided through the toy activity, for example, the coconut tree, the component of the coconut and the use of coconut, and its widespread use in local agriculture, medicine and traditional food.

Conclusion: bi-gnosis and border crossings

Braund and Reiss maintain that:

> Learning is a process of active engagement with experience. It is what people do when they want to make sense of the world. It may involve the development or deepening of skills, knowledge, understanding, awareness, values, ideas and feelings or increase in the capacity to reflect. Effective learning leads to change, development and the desire to learn more.
>
> (Braund and Reiss 2004: 5)

The aim of our work is not just to encourage one form of understanding, but two. To foster bi-gnosis, we want participants to gain information about scientific knowledge (western science) and local Thai wisdom, equally and in parallel. Our position is similar to that of Aikenhead (2001), who stated that Aboriginal students should learn western science but without being assimilated into western culture, that is, without losing their cultural identity as Aboriginals. That is, bi-gnosic learning means having access to both scientific knowledge and local wisdom as distinctive knowledge systems, each of value in its own right, and not some admix, integration or infusion of the two. Our position, then, is that rather, than seeing traditional toys as a step *towards* WMS, we see the two knowledge systems sitting alongside each other, in balance and making distinctive contributions to understanding.

There are two profound limitations to western scientific knowledge. First, science is limited by its focus on selected attributes to the exclusion of others. This is a choice made by scientists not a limitation imposed upon science by physical reality. Second, the advancement of science and science education often competes with national interest in maintaining the integrity of traditional culture. However, it would be a mistake to infer that our discussion is 'soft' on superstition, or that science is being reduced to an aspect of cultural relativism. In this kind of work, it is necessary to ask about the balance between these two interests: science and local culture. While we are encouraging bi-gnosis, the degree or kind of bi-gnosis is clearly affected by characteristics in the two systems and the relationship between them. We do not advocate the 'infusion', 'assimilation' or 'incorporation' of one knowledge system within the other, nor the 'integration' or 'blending' of the two knowledge systems. We do not support a 'middle path', a consensus approach: working away until something mutually acceptable is found.

A strategy of bi-gnosis emphasises that each culture has different content and values, and that a person can operate in both cultures. Being bi-gnostic is being able to see the same situation through both Thai and western eyes, then having the choice of which way to respond. Having more than one form of knowledge is a personal and community resource that gives more choices in action and thought, therefore more freedom. Thai local wisdom is a way of life, knowledge that tries to make people aware of life and adapt themselves to nature. For example, the coconut mouse toy is made from coconut shell, a wooden wheel, string and an elastic band. These materials are 'close to nature', can be found easily, are hand-made, and construction is straightforward. This toy-making encourages adult and children's creativity so that, when they make and decorate a coconut mouse toy, they often create other animals, for example, coconut turtles, coconut beetles, coconut cars etc.

Bi-gnosis seeks understanding from both sides of the issue, deriving two a clear understandings from both Thai local wisdom and Western Modern Science that can work side-by-side. We tend towards Article 8J of the UN Convention on Biodiversity, which reads:

> ... respect, preserve, and maintain knowledge, innovations, and practices of Indigenous communities and lifestyles relevant for the sustainable use of biological diversity. And promote their wider application with the approval of such holders of such practices. Encourage the equitable sharing of benefits from the uses of such knowledge and practices.

In this way, we do favour the sense of a teacher being a 'cultural broker'. This term has been used to describe ways in which teachers facilitate student 'border crossings' between LIS and WMS. For example, Aikenhead describes the ways in which cultural brokers act:

> They acknowledge that a border exists, they motivate students to cross it, they employ language of both the students' culture and the culture of western science, they explicitly keep track of which culture comprises the context at the moment, and they help students resolve cultural conflicts that may arise.
>
> (Aikenhead 2006: 121)

Cultural brokers motivate students by understanding the specific history of the students' culture, and by maintaining high expectations.

Key questions

Given what has been said above, that – ideally – both knowledge systems ('everyday' and WMS) do not 'infuse', 'assimilate', or 'integrate', then:

1   What is relationship of both forms of knowledge, or is there no relationship?
2   If they do have relationship together, how do both of them stand, stay or live with each other?

Further reading

Kanhadilok, P., and Watts, D. M. (2012) Family play-learning: Some learning outcomes from make-and-play activities with toys at a science museum. *Literacy Information and Computer Education Journal*, Special Issue, 1(2): 879–85. This gives some broad background to the issues I discuss here.

Kanhadilok, P., and Watts, D. M. (2013) Western science and Thai local wisdom: Using museum toys to develop bi-gnosis. *Canadian Journal of Science Mathematics and Technology Education*, 13 (1): 33–48. While you read this paper, there are several others in the same issue that are worth reading too.

Aikenhead, G. (2001) Integrating Western and Aboriginal sciences: Cross-cultural science teaching. *Research in Science Education*, 31: 337–55. Glen Aikenhead is the 'guru' in many of these matters. Anything he writes is worth reading, hence I also include the following.

Aikenhead, G. (2006) *Science Education for Everyday Life: Evidence-based practice.* London: The Althouse Press.

Akenhead, G. (2011) *Cross-culture science teaching: Praxis.* A paper presented at the annual meeting of the National Association for Research in Science Teaching, St. Louis, 26–28 March.

Akenhead, G., and Ogawa, M. (2007) Indigenous knowledge and science revisited. *Culture Study and Science Education*, 2: 539–620.

References

Ackoff, R. L. (1989) From data to wisdom. *Journal of Applied Systems Analysis*, 16 (1): 3–9.

AHRC (Arts and Humanities Research Council) (2012) *Science in culture, large grant funding*. Swindon, UK: Arts and Humanities Research Council.

Aikenhead, G. (2001) Integrating Western and Aboriginal sciences: Cross-cultural science teaching. *Research in Science Education*, 31: 337–55.

Aikenhead, G. (2006) *Science Education for Everyday Life: Evidence-based practice*. London: The Althouse Press.

Alsop, S. (ed.) (2005) *Beyond Cartesian dualism: Encountering affect in the teaching and learning of science*. Dordrecht: Kluwer Press.

Battiste, M. (2000) *Aboriginal Education: The post-colonial challenge after the Royal Commission on Aboriginal Peoples*. Plenary, Education as Life-long Learning Conference, University of Saskatchewan, November 24.

Battiste, M., and Henderson, J. (2000) *Protecting Indigenous Knowledge and Heritage*. Saskatoon: Purich Publishing.

Boonlert, P. (2012) *Hello the mouse year: The first of the animal year cycle*. Available online at: http://campus.sanook.com (accessed 22 March 2011).

Braund, M., and Reiss, M. (2004) *Learning Science Outside the Classroom*. New York: RoutledgeFalmer.

Cajete, G. (2000) *Native Science: Natural laws of interdependence*. Santa Fe, NM: Clear Light Publishers.

Carruthers (2011) *Hands-on Learning*. Available online at: http://www.butterflyfield.com/index.php (accessed 9 October 2011).

Garroutte, E. (1999) Getting serious about 'interrogating representation': An 'indigenous turn'. *Social Studies of Science*, 29(5): 589–600.

Hammond, L., and Brandt, C. (2004) Science and cultural process: Defining an anthropological approach to science education. *Studies in Science Education*, 40: 1–47.

Kanhadilok, P. (2011) 'Technical Toys and Playthings: Informal Science Education in Science Museums.' The 6th World Congress of the International Toy Research Association, 26 and 30 July, Bursa, Turkey.

Kawagley, O. (2000) Identity-creating camps. *Sharing Our Pathways*, 5(2): 4–5.

Kongprasertamorn, K. (2007) Local environmental protection and community development: The clam farmers in Tambon Bangkhunsai, Phetchaburi Province, Thailand. *Manusya: Journal of Humanities*, (10)1: 1–10.

McKinley, E. (2005) Locating the global: Culture, language, and science education for indigenous students. *International Journal of Science Education*, 27: 227–41.

Michell, H. (2005) Nehithawak of Reindeer Lake, Canada: Worldview, Epistemology and relationships with the natural world. *The Australian Journal of Indigenous Education*, 34: 33–43.

Michel, H., and Gayton, D. (eds) (2002) *Linking Indigenous Peoples' Knowledge and Western Science in Natural Resource Management: Conference proceedings*. Kamloops, BC: Southern Interior Forest Extension and Research Partnership. SIFERP Series 4.

Na Talang, A., Pongpit, S., and Nakabut, A. (1995) *Wisdom and Learning Process of Thai Local People: The Kittimatee Project of Education*. Bangkok: Sukhothaitummatirat University Press.

National Science Museum (2007) *Traditional Technology*. Pathumthani, Thailand: National Science Museum Press.

Nearnchalearm, P. (2005) Learning science through local toys. *Wichakarn Journal*, 8(4): 17–24.

Office of SMEs Promotion (2010) *Stories of Thai Local Wisdom*. Bangkok: Office of SMEs Promotion (OSMEP).

Office of the National Education Commission (1999) *The Ways to Encourage Thai Local Wisdom into the Education System*. Bangkok: Primdee Printing.

Office of the National Education Commission (2011) *Indigenous Knowledge for a Learning Society*. Bangkok: Office of the National Education Commission.

Ogawa, M. (1995) Science education in a multi-science perspective. *Science Education*, 79: 583–93.

OSMEP (Office of Small and Medium Enterprise Promotion) (2012) *One Tambon, One Product (OTOP): Stories from Thai local wisdom*. Bangkok: Office of Small and Medium Enterprise Promotion.

Pongpit, S. (1994) *Local wisdom and the development of community*. Bangkok: Amarin Printing Group.

Pungkanon, K. (1992) The variety of views about 'Local Wisdom'. *The National Education Journal*, (24), September–October: 36.

Ritchie, S., and Butler, J. (1990) Aboriginal Studies and the science curriculum: Affective outcomes from a curriculum intervention. *Research in Science Education*, 20: 249–354.

Seri, P., and Nantasuwan, W. (2002) *Master Community Plan: Sustainable development*. Bangkok: Charoenwit Printing.

Snively, G., and Corsiglia, J. (2001) Discovering indigenous science: Implications for science education. *Science Education*, 85(1): 6–34.

Stephens, S. (2000) *Handbook for Culturally Responsive Science Curriculum*. Fairbanks, Alaska: Alaska Native Knowledge Network.

Sungsri, S. (2011) *The Role of Local Wisdom in Promoting Lifelong Learning in Thailand*. Bangkok: Office of Educational Services Press.

Taylor, N., Littledyke, M., Eames, D., and Coll, R. K. (2009) *Environmental Education in Context: An international perspective on the development of environmental education*. Netherlands: Sense Publishers.

Tinnaluck, Y. (2005) *Knowledge Creation and Sustainable Development: A cllaborative process between Thai local wisdom and modern sciences*. PhD Thesis. Paris: University of Paris.

Yoksan, S. (2011) *The story about the mouse*. Bangkok: Thaiwattanapanit Press.

Part V

Postscript

Chapter 17

Decisions and time to take sides

Mike Watts

To end the book as I began, on our way to a family wedding last summer, we stopped overnight at a small chateau in France. Over a good dinner that night, we four of us fell to arguing about our 'three most important technological inventions that have changed the world'. It is a favourite theme, an endless source of debate. Oscar delighted in disparaging my three: soap, glass and the safety match. Too prosaic by far. He sided rather more with Ruth's mobile phones, television and champagne, and scorned Rosie's pigments and dyes, fabrics and photography. Oscar's own were the Internet, the internal combustion engine (he's a self-proclaimed addict) and Chelsea football team (ditto), which just about sums him up perfectly. Apart from what our choices say about our personal preferences, the discussion itself was remarkable for the heat and emotion we generated. There is *no way* that Chelsea football can be classed as a technological invention, except when you come up against Oscar's resilient and determined arguments. And have dyes and fabrics really changed the world? They have if you listen to Rosie. Ruth made her own point rather nicely by tackling the case of and for champagne with gusto. She lost track of much of the conversation after that. We were so wrapped up in the heat of the moment we were slow to realise that the dining room had emptied of the more genteel French diners around us, or maybe because of us.

The key aspect here is that debate is seldom a rational and logical matter. Nor, for that matter, is science, even less so technology. When people talk about science there is often an explicit – more likely implicit – notion of some hallowed principles that are called 'the scientific method'. Leave aside for a moment that this is a highly-contested set of ideas, and that there are many people who argue that no such 'method' exists. Let us imagine, instead, that there are some essential aspects of science that lend to it both insight and authority. These are usually couched in terms of observation, hypothesis, experimentation through control of variables, measurement, analysis of outcomes, and some discussion of interpretation and implications. So far so good.

However, science uses many different forms of logic, and each is relevant to the development of scientific thinking in both children and adults. For example, there is the common forms of rational mathematical-logical reasoning such as 'if this … then that …' or '… not that'. This is close to what we mean by logical thinking; it is at the heart of causality.

Question: What is the difference between force and pressure in physics?

Answer: We define forces as acting on a body *through a point of application*; pressure is an effect that occurs when a force is applied on a surface, it is the amount of force acting on a unit area of a body. One difference lies in terms of mathematical distinctions and the notion of a Euclidean point: in Euclidian geometry, a point has no dimensions and, therefore, no area. Therefore, force is an idealised notion, where real-world pressures are deemed to be focused down to a dimensionless point, so that we can then make the handling of the computations (the mathematics) easier. Force is a vector, and so we use the mathematics of vector algebra to solve force-related problems. On the other hand, pressure is neither a scalar nor a vector, it is another mathematical aspect altogether, a tensor. Pressure has no direction, it is transmitted to solid boundaries or across arbitrary sections of a fluid normal to these boundaries or sections at every point. It is a fundamental parameter in thermodynamics and it is conjugate to volume. In this sense, pressure is *cause* and the force is *effect*.

So, simply understanding that last paragraph entails a wide set of logical underpinnings to what many feel are easy 'everyday' and 'school' concepts such as force and pressure.

There is also the notion of variance and invariance, what we know to change and not change easily. For an invention to have an enduring impact rests upon the epochs and eras in question, the extent to which something remains constant over time. Moreover, we both interpolate and extrapolate from data so that, while we might fix two points, we can also look at moments in between, and then off into the distance. Photography is a relatively recent invention and is already passé (do you count digital images as photo 'graphs'?). And then Oscar's favourite device, *reductio ad absurdum*. This is a common form of argument that looks to demonstrate that a statement is true (or false) by showing that a stupid, untenable or absurd result then follows.

> *Reductio ad absurdum*, which Euclid loved so much, is one of a mathematician's finest weapons. It is a far finer gambit than any chess gambit: a chess player may offer the sacrifice of a pawn or even a piece, but a mathematician offers the game.
>
> (Hardy 1993: 34)

Other forms of logic surround correlations and probabilistic connectivity: what seems reasonable, what is a likelihood, what is certainty and uncertainty. There's case-based reasoning (the way we use cases as the basis of argument; the use of analogy and metaphor, evidential reasoning). Beyond that, there is narrative logic (the coherence and fidelity of stories; the precedence of meaning over logic; the importance of quality and character; the sense of processes and consequences). So, while common mathematico-logical reasoning does underpin the development of scientific thinking, it is not the only form of logic at play by any means.

Take, for example, 'quoughts'. These are thought questions, a term that has crept into the literature on learners' questioning to capture 'thought puzzles', unspoken perplexities, internal moments of wonderment (Pedrosa-de-Jesus and Watts 2011). Thought experiments are famous throughout science and the philosophy of science, and have received detailed attention, not least through Karl Popper's (1968) *On the Use and Misuse of Imaginary Experiments, Especially in Quantum Theory*, and Thomas Kuhn's (1977) *A Function for Thought Experiments*. There is a long list of thought experiments: Borel's Monkeys; Einstein's Elevator; Einstein's Light-beam; Hawking's Turtles; Heisenberg's Gamma-ray microscope; Mach's 'Welcome to the edge of the universe'; Maxwell's Demon; Newton's Bucket; Schrödinger's Cat; Searle's Room; Parfit's Teleporter; Putnam's Twin-Earth, to note just a few.

One favourite is Galileo and the Tower of Pisa. Contrary to popular myth, Galileo Galilei did not drop balls from the Tower of Pisa; he asked the question and conducted the gravity experiment in the 'laboratory of his mind'. His sixteenth-century peers believed heavier objects fell faster than light ones. So Galileo imagined a heavy ball attached by a string to a light ball. His question was: 'Would the light ball create drag and slow the heavy one down?' He thought that was absurd, reducing his opponents' arguments by ridicule (and failing to endear himself to them). No, he concluded, the balls would hit the ground simultaneously.

So, have we contributors to this book generated debate? It is to be hoped so. Will readers have been challenged in their thinking, have been required to do some re-thinking? Again, that is one of the anticipated outcomes. Will they have chosen sides or, at least, sided with arguments? Another hoped for product of reading into a book like this.

Meanwhile, have we exhausted science's controversial issues? Not by a long chalk. There are many more, not least controversies over science and religion, science and maths, science and society, science and superstition – to name but a few. But those are all debates for another day.

References

Hardy, G. H. (1993) *A Mathematician's Apology*. Reprinted with a foreword by C. P. Snow. New York: Cambridge University Press.

Kuhn, T. (1977) A function for thought experiments. Reprinted in T. Kuhn, *The Essential Tension*. Chicago: University of Chicago Press (pp. 240–65).

Pedrosa-de-Jesus, M. H., and Watts, D. M. (2011) Questions and science. In R. Toplis (ed.) *How Science Works: Exploring effective pedagogy and practice*. London: Routledge.

Popper, K. (1968) On the use and misuse of imaginary experiments, especially in Quantum Theory. In K. Popper, *The Logic of Scientific Discovery*. London: Routledge (pp. 442–56).

Index